Elio Schaechter
Artwork by Judith Schaechter

In the Company of Microbes

Ten Years of *Small Things Considered*

Washington, DC

Library of Congress Cataloging-in-Publication Data

Names: Schaechter, Moselio, editor. | Schaechter, Judith, 1961- illustrator.
Title: In the company of microbes : ten years of Small Things Considered /
edited by Elio Schaechter ; artwork by Judith Schaechter.
Description: Washington, DC : ASM Press, [2016] | ?2016 |
Includes bibliographical references.
Identifiers: LCCN 2016011177 | ISBN 9781555819590 (pbk.)
Subjects: LCSH: Microbiology–Blogs. | Microorganisms–Blogs.
Classification: LCC QR56 .I5 2016 | DDC 576--dc23 LC record available at
http://lccn.loc.gov/2016011177

10 9 8 7 6 5 4 3 2

Printed in the United States of America

Address editorial correspondence to ASM Press, 1752 N St., N.W., Washington, DC 20036-2904, USA

Send orders to ASM Press, P.O. Box 605, Herndon, VA 20172, USA
Phone: 800-546-2416; 703-661-1593; Fax: 703-661-1501
E-mail: books@asmusa.org
Online: http://estore.asm.org

Cover image: Judith Schaechter
Design: Lou Moriconi

In the Company of Microbes

Ten Years of
Small Things Considered

Table of Contents

Part 1: The View from Here

Part 2: Accounts of the Past

Part 4: On Being a Microbiologist

Part 5: Personal Notes

Part 6: The Ways of Microbes

Part 7: Teaching Things

Contributors

Bonnie L. Bassler
Princeton University, Princeton, NJ

John Coffin
Tufts University, Medford, MA

Chris Condayan
American Society for Microbiology, Washington, DC

Julian Davies
University of British Columbia, Vancouver, BC, Canada

Tim Donohue
University of Wisconsin, Madison, WI

Stanley Falkow
Stanford University School of Medicine, Stanford, CA

Peter Geoghan
American Society for Microbiology Washington, DC

Daniel P. Haeusser
Canisius College, Buffalo, NY

Franklin M. Harold
University of Washington Health Sciences Center, Seattle, WA

Jaimie Henzy
Boston College, Chestnut Hill, MA

Welkin Johnson
Boston College, Chestnut Hill, MA

Suckjoon Jun
University of California, San Diego, La Jolla, CA

Patrick Keeling
University of British Columbia, Vancouver, BC, Canada

Richard Losick
Harvard University, Cambridge, MA

Phoebe Lostroh
Colorado College, Colorado Springs, CO

Mark Lyte
School of Pharmacy, Texas Tech University Health Sciences Center, Amarillo, TX

Joe Mahaffy
San Diego State University, San Diego, CA

Mark Martin
University of Puget Sound, Tacoma, WA

Jeff F. Miller
University of California, Los Angeles, Los Angeles, CA

Nanne Nanninga
Swammerdam Institute for Life Sciences, University of Amsterdam, The Netherlands

Fred Neidhardt
University of Michigan Medical School, Ann Arbor, MI

Maureen O'Malley
University of Sydney, New South Wales, Australia

Mercé Piqueras
Freelance science writer and science editor, Barcelona, Spain

Vincent Racaniello
Columbia University, New York, NY

Shmuel Razin
Hebrew University-Hadassah Medical School, Jerusalem, Israel

Gemma Reguera
Michigan State University East Lansing, MI

Elio Schaechter
University of California, San Diego, La Jolla, CA, and San Diego State University, San Diego, CA

Claudio Schazzocchio
Imperial College London, London, England

Jan Spitzer
Mallard Creek Polymers, Charlotte, NC

William C. Summers
Yale University School of Medicine, New Haven, CT

Amy Cheng Vollmer
Swarthmore College, Swarthmore, PA

Christoph Weigel
Hochschule für Technik und Wirtschaft, Berlin

Conrad Woldringh
Swammerdam Institute for Life Sciences, University of Amsterdam, The Netherlands

Charles Yanofsky
Stanford University, Stanford, CA

Introduction

Sometime early during 2006, ASM staff members and I began discussing ways to expand the society's communications efforts into a new area. We wanted to take better advantage of available electronic means for reaching ASM members and interested members of the public with information about microbiology. One opportunity was for ASM to launch a blog. However, blogging raised eyebrows in some circles, including the attorneys who are asked to review proposed programs of this sort. Being good lawyers, they mentioned liability issues, worrying that such an informal vehicle for communicating might prove too free-wheeling if not handled properly. Without a candidate blogger in mind, we pondered where to go next.

Meanwhile, Elio Schaechter contacted me out of the blue, saying that, in retirement, he was seeking a new project and wondered whether ASM might be interested in supporting a blog on microbiology. Instantly, our concerns over blogging liabilities vanished. Here was an eminent microbiologist and former president of ASM who was respected across the broad community of microbiologists. Don't worry, he assured us, the nooks and crannies of microbiology can provide plenty of rich material for the blog and, while it might touch on controversy, any forthcoming debates will be strictly Talmudic—(The Talmud is a combination of texts explaining the meaning of events or practices appearing in the old Testament)—not damaging to any institution or person, apart perhaps from those with especially fragile egos.

By then ASM had begun producing podcasts, and the blog that Elio conceived of doing fell under that umbrella—or, rather, onto that platform. Elio was introduced to Chris Condayan, a public communications manager at ASM, who says that Schaechter quickly proved to be a "sharp tech," meaning he soon mastered the virtual mechanics as well as the art of the blogging process. From the outset, Elio issued blog posts regularly, typically posting two items each week. He also very much molded its content, mining gems excavated from the broad expanse of the microbiological sciences.

To begin with, *Small Things Considered* was purely Elio's output, but eventually he attracted other writers and microbiologists to join him in this novel communications enterprise. The first additional steady contributor was Merry Youle, a technical writer, who approached him as an interested reader. For several years she helped to edit the postings before she began to write some of them herself. Later, several microbiologists were brought on board to broaden the scope of the blog postings but also to give Elio some hands-on help.

There was a flurry of interest after the first few postings during the second half of 2006, which subsided for a while before it gradually but steadily began building into a huge success. Over the past decade, Elio and his band of bloggers posted more than 1,000 items on *Small Things Considered*, and those postings by now have attracted more than 2 million page views at an average of more than 600 views per day, and elicited more than 2,300 comments. Along with traditional postings, the blog now includes a Teachers Corner that caters to classroom needs. Among microbiologists, the blog is well known and much liked, but it also attracts plenty of readers outside the discipline. Further, it is so well respected that the Library of Congress has identified *Small Things Considered* as one of the first blogs in the sciences worth archiving.

The blog could only have been conceived by Elio. His store of knowledge about the science of microbiology is unequalled. He is also an unusually articulate writer and has a unique and welcoming approach to the content.

The 70 articles are divided into seven sections entitled: "The View from Here," "Accounts of the Past," "Small Wonders," "On Being a Microbiologist," "Personal Notes," "The Ways of Microbes," and "Teaching Things"

The authors, in addition to Elio, are presidents of ASM and a slew of distinguished microbiologists from all around the globe. The topics are as varied as the authors, ranging from, "Where Mathematicians and Biologists Meet" to "Bacterial Hopanoids: The Lipids That Last Forever."

This is more than just a collection of articles: it is a treasure chest of wise, amusing, and even profound statements about the ubiquity and relevance of the microbial world. As Elio notes in his introduction, "The purpose of this blog is to share my appreciation for the width and depth of the microbial activities on this planet. I will emphasize the unusual and the unexpected phenomena for which I have a special fascination."

Michael Goldberg, PhD
Executive Director Emeritus, ASM

Jeff Fox PhD
Current Topics and Feature Editor, *Microbe*

Preface

Blog Years Are Like Dog Years

How did *Small Things Considered* come about? When I retired from Tufts University some twenty years ago, I realized that for too long I had lived comfortably within the focused world of *E. coli* and its ilk. Out there, far from the lab bench, amazing and unexpected things were happening, and I was only dimly aware of some. A few of them danced in my imagination; many were just out of sight. So, I started to pay more attention to these groovy stories.

My itch of writing egged me on to look for a way to share my pleasure of exploration with others. There weren't many venues for such endeavors then, but someone whispered blog in my ear. I barely knew how to spell it, leave alone what it was, so I called Michael Goldberg, the then Executive Director of the the American Society for Microbiology (ASM), who encouraged me and put me in touch with Chris Condayan, who knew a lot about such things. Not only did Michael explain this enigmatic word to me, but he also told me that they had been talking about an ASM-sponsored blog, and, yes, if I were willing to do this, they would help me out. Chris set me up by designing the lay-out, plugging me into the proper software, and even coming up with a happy name for the blog. As an aside, when I was once interviewed on the NPR radio program *All Things Considered*, I coyly asked them if they minded our using a name derived from theirs. They said that they were actually pleased.

This blog began in 2006. Blog years are indeed like dog years, and a decade is a long time. This milestone is worth celebrating, I think. We (and I will explain the "we" right away) have been at it assiduously for all this time and, miraculously, have not missed a single one of the scheduled postings. These have been bi-weekly, Monday being devoted to longer items, Thursday to brief ones such as Talmudic Questions, Pictures Considered, or Terms of Biology. Sometime last September we reached 1000 posts, and we still doggedly continue to produce them at our habitual pace.

The "we" refers to the team that has been working with me for much of this time. Merry Youle joined me in this effort almost immediately after its inception. In the course of time, we became partners and for several years shared the responsibility for writing most of the posts, pitilessly editing each other's drafts. From this emerged a comradely friendship that continues to this day. In time, we asked others to join us, making for an interactive and productive team. The current members are Daniel Haeusser, Jamie Henzy, Gemma Reguera, and Christoph Weigel. I must mention Marvin Friedman, who would still be contributing a post a month were he still with us. Early on, I opened the pages of this blog to others, especially graduate students, who, I reckoned, could benefit from the experience. I am glad that blogging by students has become a widespread activity elsewhere. Others, including notable people in the field, have presented their thoughts and opinions in STC. Nowadays most of the blog items are indeed authored by folks outside our inner group, and the choice of articles for this book reflects that.

This book was started by a prompt from Chris. For this collection I scanned our archives and focused on material that one could broadly call musings—reflections on personal and historical interactions between the writers and microbes. A few of the pieces included fall outside this scope and are mostly about unusual discoveries. Why did I not opt for more science-oriented stuff? Two reasons: one is that their appeal would be narrower; the other is that such material becomes rapidly dated. And besides, wouldn't you want to hear what questions and puzzles still animate microbiologists, what they think about both the past and the future of the field?

Our 1000 posts have garnered some two million views. I guess that's something to brag about, but more to the point is the enjoyment that this effort has brought me. I have always had something of a naturalistic bone in my body. Being obliged to work at the bench, pleasurable though it has been, kept me from relishing the small wonders that are "out there," where there is a never-ending pageant of astounding variations on the theme of microbial life. In old age, such hankerings are to be indulged and such wanderings to be treasured. They help provide answers to some of the eternal questions that I (and you too, dear reader) have been asking intermittently since adolescence. So, my deepest thanks to all who have given me this jewel of an opportunity but most of all, to the small things that are waiting to be considered.

Elio Schaechter

October 2015

San Diego, California, USA

Acknowledgments

My thanks go to all who have contributed to the blog over the years and have helped make exciting stories accessible to a wide public. I single out Merry Youle, who was my partner for several years and a friend. She helped me in every way imaginable, from the choice of topics, to doing the wisest of editing, to providing me with a moral compass, to holding my hand (at a distance, as she lives in Hawaii, I in California). I thank my other collaborators on this endeavor: Gemma Reguera, Daniel Haussler, Jamie Henzy, and, early on, Mark Martin and Welkin Johnson. Christoph Weigel needs to be acknowledged in a special way for his extraordinary contributions to both the blog's content and format. Also, I thank the late Marvin Friedman who, for many years, was another faithful contributor. All these persons did more than supply material: they participated in thinking about ways this blog could better fulfill its purpose. They have all done it with verve, passion, and great insights. I have been very lucky indeed.

I thank the American Society for Microbiology (ASM) for sponsoring the blog, especially the Communications staff directed by Erika Shugart. Chris Condayan generously and imaginatively supported this effort from the very beginning and has been a source of help and encouragement ever since. Andréa Gwartney did a superb job dealing with the layout and organization of the material. Ray Ortega later took charge of producing each issue of the blog, using a keen eye and warm heart in order to achieve this blog's intended purposes. All of these persons have treated their work as a labor of love.

I thank my wife, Edith, for her support. She is not a scientist but has played a huge role in everything else it takes to carry out such an activity. I also thank my daughter Judith for allowing me to use her drawings. She is a renowned stained glass artist whose doodles are often of imaginary protists and diatoms. I never expected to see both our names on the cover of the same book.

What do we mean by a Talmudic Question?

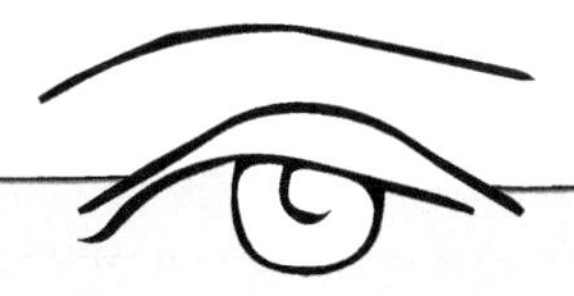

The term is borrowed, loosely and perhaps inappropriately, to describe questions whose answers cannot be found by a Google search. In most cases, the questions don't have ready answers but are intended to provoke thought and discussion. We don't aim to be disrespectful of the old and venerable tradition of the Talmudic Method.

Old TQs never expire. They remain open for you to add your own response. A link or URL is provided for each TQ in the book, or you can search for them by number, e.g., "question #73" for TQ #73. Warning! Once started, you may be tempted to explore the many TQs left, by necessity, out of the book.

PART 1

The View from Here

I invited experienced microbiologists to share some of their musings, and I added a few of my own.

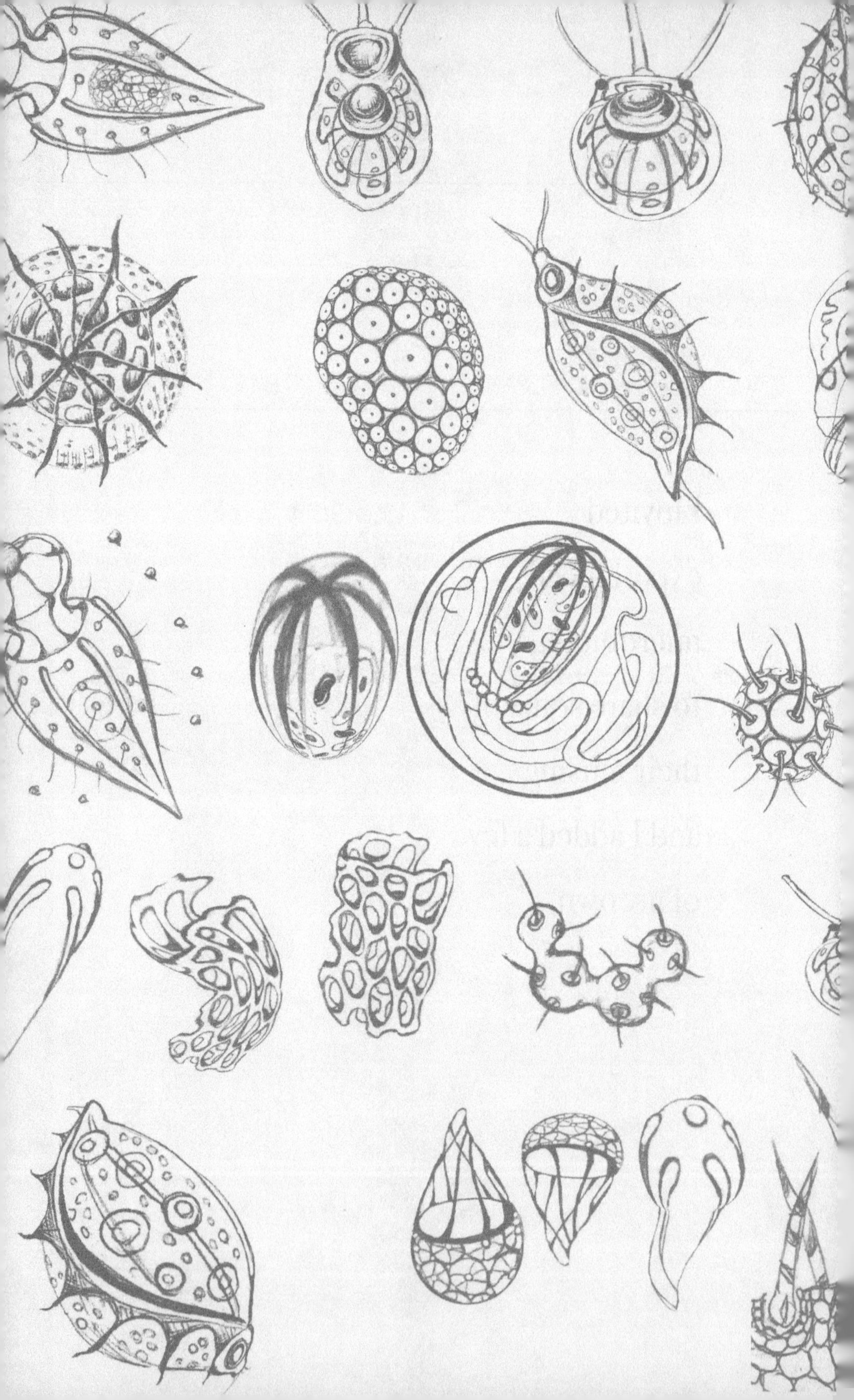

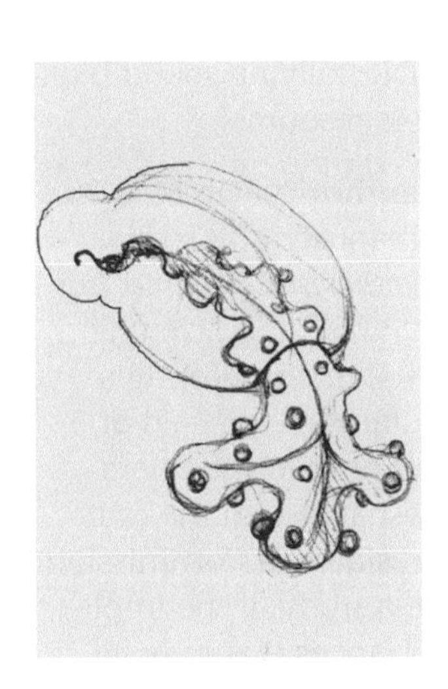

1

Of Ancient Curses, Microbes, and the ASM

by Bonnie L. Bassler

On July 1, 2010, when I started my term as ASM president, I was reminded of three ominous curses of dubious ancient origin:

1. May you live in interesting times.
2. May you come to the attention of those in authority.
3. May you find what you are seeking.

May you live in interesting times: Clearly, these times qualify. Microbes will be at the heart of solutions to our most pressing problems: the environment, food, energy, and health. The BP oil spill began on day −79 of my term. Microbes are coming to the rescue, and ASM expertise is on the scene Let us hope that lessons have been learned. In his inauguration address, President Obama promised to "*restore science to its rightful place*." It may be happening. The National Academy of Science's Board on Life Sciences recently released its report: *A New Biology for the Twenty-First Century*. The US Cabinet Secretaries of Energy and Agriculture requested a series of workshops to discuss how to implement the new biology. A first workshop, focused on food and fuel, was held in DC last month, and I was invited to participate (day −27 of my term). Workshop members were asked to develop scientific challenges for the decade to be proposed to Congress for funding. Together with several other ASM members in attendance, I strongly advocated the understanding and appropriate use of microbes to synthesize new fuels, clean up the environment, optimize crop production, etc. Our rallying sound bite: microbes, the world's only unlimited renewable resource!

May you come to the attention of those in authority: On May 20, with quite some fanfare, *Science* published a manuscript: *Creation of a Bacterial Cell Controlled by a Chemically Synthesized Genome*. On that same day (day –42 of my term) the ASM Public and Scientific Advisory Board (PSAB) released a position statement offering a balanced perspective on the manuscript, the status of the field of synthetic biology, and its regulation. A number of ASM members, including myself, gave expert opinions in newspapers, on the radio, and on TV. President Obama requested his Presidential Commission for the Study of Bioethical Issues to undertake a study of the implications of this scientific research and other advances that may lie ahead in this field. I will testify at the first Commission hearing on July 8 (day +8 of my term—at last we're into positive numbers!). I feel well equipped to represent us. Last year I was the organizer and chair of the National Academies of Sciences Keck Futures Conference on Synthetic Biology. I continue to learn about new developments in the field, and I am expertly advised by our PSAB staff, our new PSAB Chair Roberto Kolter, and other knowledgeable ASM members. My specific role is to compare and contrast the engineering perspective with that of the biological and genetic sciences, and to explain how approaches represented by synthetic biology differ from other approaches to biological manipulation. I will also address what has been accomplished and what is likely to be accomplished in this field and what I think are important obstacles to the advancement of synthetic biology.

May you find what you are seeking: Over this past year (day –365 to day 0) as president-elect, I got to know the Society inside and out. I learned what remarkable accomplishments 40,000 volunteers can achieve. I saw our members donate huge blocks of time for the good of our discipline and the health of our planet. I became fully convinced that collectively we have the potential and the expertise to make the world healthier, to enable a sustainable relationship with our environment, and to ensure the promise and prominence of science and technology in our culture. It has been a magical and eye-opening year. I am beginning to understand the vast breadth and depth of this organization and the position of its members in leading the nation and the world in all in matters touched by microbes.

There's nothing like a few ominous curses to get the blood flowing. I am eagerly looking forward to this coming year with the hope that I can make significant contributions to the ASM and to our community at large. I will do my best to further enhance our reputation in public policy, education, outreach, and scientific advancement. Now at day + 1 and counting, I look forward to meeting you!

We have a tradition of hosting a few reflections from the incoming president of the ASM.

Bonnie Bassler is the Squibb Professor of Molecular Biology at Princeton University and a Howard Hughes Medical Institute Investigator. She is a fellow of the American Academy of Microbiology and a member of both the National Academy of Sciences and the American Academy of Arts and Sciences.

July 1, 2010

bit.ly/1LAnyRd

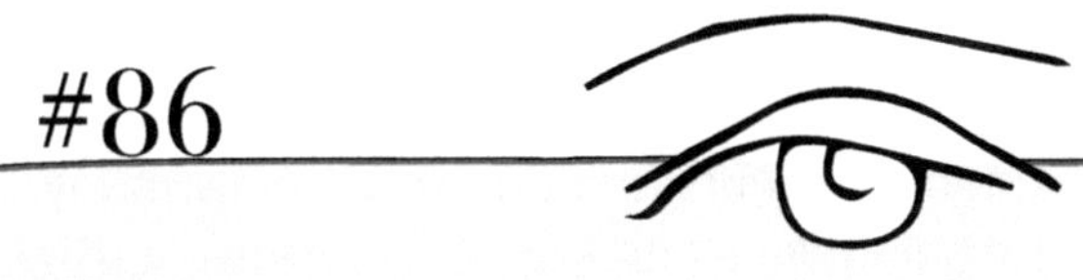

by Elio

Given that so many kinds of bacteria are intimately associated with animals and plants, why are so relatively few pathogenic?

April 12, 2012

bit.ly/1NRFuN5

2

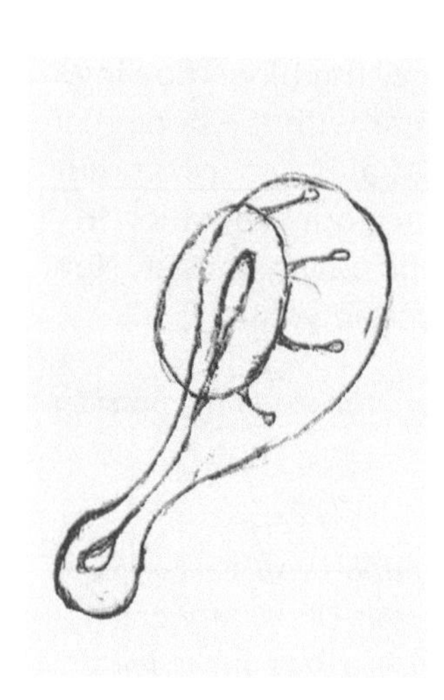

The View from Here: The Evolution of the Genetic Code

by Charles Yanofsky

Although the genetic code is well established, a very exciting unsolved problem is discovering how codons were related to amino acids in the evolution of protein synthesis. How did a tRNA and a tRNA synthetase first evolve, and what was their ancestral source? How were the genes for the first tRNA and tRNA synthetase duplicated, and how was their specificity varied? Can we offer any explanation for why there are two classes of tRNA synthetases? Can one predict which tRNAs evolved from one another? Similarly, can we predict which tRNA synthetases evolved from an existing tRNA synthetase?

A related series of exciting experiments would be to attempt to reproduce some evolutionary events and determine how many mutational changes it would take—and where—to evolve a tRNA synthetase with new specificity from an existing synthetase. Suppression studies many years ago showed that tRNAs may acquire new decoding specificity following single mutational changes, but synthetase suppressors were not recovered, as I recall. Also, we now know that synthetases recognize both the anticodon and acceptor end sequence of each tRNA. Is this consistent with what is known about suppressing tRNAs?

Reference

Carter CW Jr. 2008. Whence the genetic code? Thawing the 'Frozen Accident'. *Heredity* **100**:339–340.

Charles Yanofsky is Professor Emeritus of Biological Sciences at Stanford University and a 2003 recipient of the National Medal of Science.

May 8, 2008

bit.ly/1Gfy8R5

#19

by Elio

You are stranded on a desert island. During a walk on the beach, you stub your toe against a bottle, which rolls against a rock and breaks. A genie is liberated, eager to grant you one wish. You ask for a microbiological laboratory fully equipped to your specifications. The genie grants you that but with the condition that you can study only one sample. What sample would you collect for study, what would you do with it, and why?

July 8, 2007

bit.ly/1LHnXVC

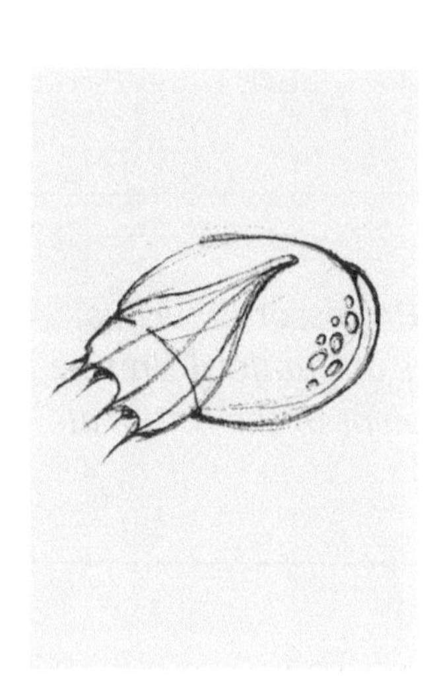

Microbes Touch Everything

By Tim Donohue

As I wheeled my bag away from the Convention Center in New Orleans after the 115th General Meeting, I noticed its ASM luggage tag, which states "*Microbes Touch Everything.*" This simple line is a perfect summary of the importance of the microbial sciences and the core mission of ASM to promote and advance this field. In history there are many examples of times when the right confluence of people and technology has led to events that rapidly push back frontiers. Two examples, one old, one of our time—Leeuwenhoek seeing animalcules through his lens and the elucidation of the structure of DNA and the genetic code—both transformed science and opened new fields of exploration. The microbial sciences are now poised to enter a renaissance that is based on interdisciplinary approaches and new technologies.

In order to take advantage of this scientific moment, we must reach out to our colleagues in other fields. At the General Meeting I heard many talks that highlighted the importance of bringing new scientific disciplines into the microbial sciences. Du-ring his opening lecture, Pieter Dorrestein shared his exciting work to understand the activities of microbial communities using mass spectrometry. He began and ended his talk by noting that he is a chemist, but that his work is in the microbial sciences. Minyoung Chun of the Kavli Foundation reflected on the foundation's interest in microbiology during the president's lecture. She noted that Kavli has historically funded researchers focused on astronomy, nanoscience, and neuroscience. In the past year they have held a series of symposia focused on the microbiome because they repeatedly heard from their physical scientists that microbiology is one of the most exciting new frontiers and thus is worth exploring. From looking for alien life to connecting the brain to the gut microbiome, microbiologists are now involved in a huge number of other fields. Clearly, our attention to interdisciplinary science must continue to grow.

Parallel to the expansion into the microbial sciences of traditionally separate fields is the boom of new technologies that are making it easier to see the unseen and contribute to understanding the world around us. For decades microbiologists were limited to studying the small fraction of the organisms that were culturable. The development of extraordinary sequencing techniques has enabled us to begin to study the other 99 percent, the microbial "dark matter" that is all around us. This technology holds great promise for the discovery of new drugs and such needed products as biofuels. Think of CRISPR-Cas as another discovery whose applications hold great promise. The ability to readily move genes into new positions is proving to be a boon to what was called the biotech industry when I was a student.

One of the most obvious areas of study that creates interdisciplinary research and new technology is that of the microbiome. Microbiologists are joining with computer scientists, ecologists, engineers, imaging experts, plus others to understand the complex ecosystems of the microbes that touch us or impact our environment in yet unknown ways. These efforts have attracted the attention of the White House, which recently sent out a request for an account of the federally-funded work in this area. Microbiome discoveries garner significant attention from the news media and the public as well.

We can work as a community to make a microbial sciences renaissance occur, or we can let the opportunity pass us by, slowing down the progress that has been made. If we become too insular, overhype our findings, and allow overly restrictive regulations to impede progress, then we have mainly ourselves to blame. As microbial sciences research becomes more complex and interdisciplinary, we must learn the best practices of team science in order to make collaborations work successfully.

The need to articulate the risk, benefits, and need for basic science was made clear by AAAS president and ex-Congressman Rush Holt in his president's forum speech on Monday. There have been calls not only from policy makers, but from microbiologists as well to look into the impact, ethics, and regulation of the use of microbes. While we may not always agree on the details, we must each participate in the ongoing dialogue about the progress that can be made in the responsible conduct of science. I call upon you to look outward to your colleagues, your community, and the ASM to explore the possibilities that the microbial sciences hold and to help push our field forward. Success in this endeavor is needed for us to truly understand how and why "*Microbes Touch Everything.*"

We continue our tradition of hosting a few reflections from presidents of the ASM.

Tim Donohue is a professor at The University of Wisconsin at Madison and a past president of the American Society for Microbiology.

July 30, 2015
bit.ly/1LleAJS

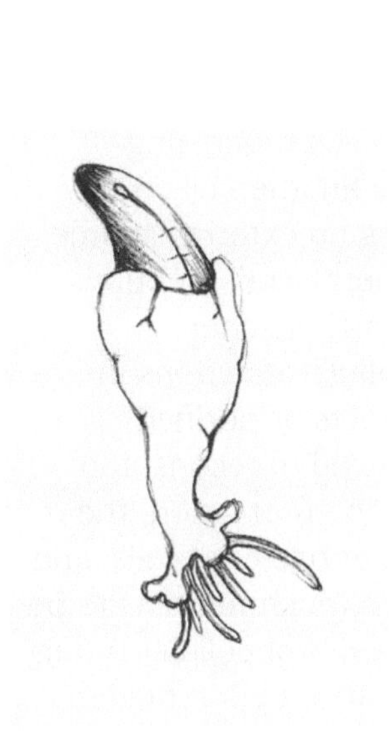

Getting a Handle on Cell Organization

by Franklin M. Harold

Structural organization is one of the most conspicuous features of cells, and possibly the most elusive. No one really doubts that cell functions commonly require that the right molecules be in the right place at the right time, or that spatial organization is what distinguishes a living cell from a soup of its molecular constituents. But the tradition that has dominated biological research for the past century mandates a focus on the molecules, and so our first step is commonly to grind the exquisite architecture of the living cell into a pulp. Few molecular scientists have asked whether anything irretrievable is lost by this brutal routine. Such questions as how molecules find their proper place in a framework of orders of greater magnitude, or how spatial order is transmitted from one generation to the next, have been largely neglected until recently.

Two current and quite excellent short reviews afford an entry into the wilderness. Eric Karsenti takes an historical approach to the role of self-organization in creating order on the cellular scale. The physical principles are arcane, but some aspects are actually quite familiar. We have known for half a century that supra-molecular complexes often arise by self-assembly, without any input of either information or energy; examples include lipid bilayer membranes, ribosomes, microtubules, S-layers, and virus particles. But the scope of self-organization has been greatly enlarged in recent years by the discovery that an array of dynamic structures can be generated in the presence of an energy source, usually ATP or GTP. The mitotic spindle of eukaryotes has been identified as a self-organizing machine; the endomembrane system may be another. Like self-assembly, self-construction (my term) requires no external source of information, but it does entail continual energy consumption. In a complementary article, Allen Liu and Daniel Fletcher survey a selection of efforts to reconstitute cellular functions in simplified systems, starting with cell-free extracts or purified proteins. Ingenious experimenters have managed to reconstitute the essentials of actin-based motility, membrane protrusion, the oscillatory system that localizes the midpoint of bacterial cells, and now also the contraction of the Z-ring. Though much remains to be learned, it is safe to conclude that the lower levels of cellular order, at least, are products of pure chemistry: they arise by interactions among the molecular constituents in ways that require the cell as a whole to supply energy and a permissive environment, but no spatial instructions.

This is excellent science, which takes us some way towards bridging the gulf between nanometer-sized molecules and cells in the range from micrometers to millimeters. It also extends the genome's reach deep into cellular structure. In a self-organizing system, the "instructions" must be wholly inherent in the molecular parts, and ultimately derive from the corresponding genes. It is the genome that specifies the architecture of the mitotic spindle, not explicitly but indirectly: the form and even functions of the spindle are implied in the structure of the spindle proteins, and in their interactions. And if the spindle can be envisaged as a creature of self-organization, why not the entire cell? Yes, indeed—but as we ascend the hierarchy of biological organization, the meaning assigned to self-organization and its underlying mechanisms undergo significant changes. Cells do not construct themselves from pre-fabricated standard parts; instead, they grow. And that mode of self-organization is not purely chemical, for it must produce parts that have biological functions, performed in the service of a larger entity that can compete and thrive in the wide world.

We are quite well informed about just how cells grow, and it is clear now that they do so by modeling themselves upon the existing structural framework, which is thereby transmitted to the next generation. To be sure, all the macromolecules involved in this sort of self-organization are gene-specified, but spatial order is not. There appear to be very few individual genes that prescribe dimensions, position, or orientation on the cellular scale. Instead, thanks to the ways in which cells grow, spatial cues are sometimes inherited in a manner quite independent of genes, (a well-established phenomenon known as "structural inheritance." The form and organization of cells thus stem from two distinct informational roots: the genomic instructions that specify the parts, and the continuity of cellular architecture that guides their placement. Specified proteins and cellular guidelines operate synergistically, reinforcing each other to generate form and organization. As Rudolf Virchow might have said, it takes a cell to make a cell.

Evidence to support such a holistic view of what happens during growth is scattered, but continues to accumulate. Let's glance at some examples. First, while many sub-cellular structures can be envisaged as products of self-construction from preformed parts, others cannot. A familiar instance is the peptidoglycan wall of bacteria, which consists of a network as large as the cell, made up of covalently-linked subunits. Enlargement during growth calls for extensive cutting, splicing, and cross-linking, even while keeping the wall physically continuous from one generation to the next. Second, even self-organizing structures must do so in a manner that ensures their correct placement in cellular space. A particularly neat example comes from recent work on the role of microtubules in cell morphogenesis of the fission yeast, *Schizosaccharomyces pombe* (reviewed by Martin). Microtubules define the poles of elongating cells by depositing there various members of the Tea complex, which in turn recruit additional factors. Cells of a certain mutant, orb6, grow as spheres even though they possess all this machinery. When, however, the mutant cells are grown in microfluidic channels that force them back into the cylindrical shape, the normal longitudinal orientation of the microtubules recovers, and so does deposition of polarity factors at the poles (Terenna et al.). Clearly, the microtubule system and cell form collaborate to organize the cell. Just how this comes about is uncertain, but we can borrow a clue from another admirable study, this one in *Bacillus subtilis*. Ramamurthi et al. found that the peripheral membrane protein SpoVM localizes to a particular patch of membrane during sporulation by recognizing its curvature; perhaps microtubule ends do likewise.

All this makes good sense, at least to me, but it reopens the question how, and indeed whether, the genome specifies cell morphology and organization. The classical conception, which has been articulated by such luminaries as August Weismann, François Jacob, and Richard Dawkins, sees the cell as a creature of its genes, and its form and functions as little more than epiphenomena. In the past, this gene-centered view of life has rested on extrapolation more than direct evidence, but confirmation has recently come by way of a remarkable paper from Craig Venter's laboratory. Lartigue et al. described the transplantation of the entire, naked genome from one species of *Mycoplasma* to another, tuning the recipient into the donor by both genetic and phenotypic criteria. I hasten to add that the rate of success was vanishingly small, one cell transformed out of 150,000; that no one yet knows just what transpires during genome transplantation; and that it remains to be seen whether it can span genera as well as species. All the same, this is surely a landmark study. If one takes its conclusions literally (which the authors were careful not to do), one must infer that a cell represents the execution of the instructions spelled in its genes, and nothing more.

On the face of it, there seems to be a glaring conflict between the geneticist's understanding of cell organization, and the physiologist's. The former insists that form and organization obey the genome's writ. The latter sees the genome as a key subroutine within the larger program of the cell, and it is the cell, not its genome, that grows, reproduces, and organizes itself. They can't both be true—or can they? Note that reproduction and heredity operate on different timescales. A growing cell relies on self-organization to transmit much of its spatial order, by mechanisms quite independent of the genetic instructions. But the genes specify the parts, and mutations commonly affect the higher levels of order; on the evolutionary timescale, it will be the genes that chiefly shape cells. Having said this, there remains a long stretch between the straightforward specification of an amino acid sequence by its corresponding sequence of nucleotides, and the devious and cryptic manner in which the genome can be said to specify the whole cell. Intellectual subtleties must not obscure the conceptual shift, from a linear chain of command to a branched and braided loop of causes and effects reverberating in a self-organizing web. The only agent capable of interpreting the *E. coli* genome as "a short rod with hemispherical caps" is the cell itself.

There is a whiff of vitalism about this view of life, even a hint of heresy. Stop now and take a deep breath, for once you begin to wonder where all this organization came from in the first place, you are headed for the blue water.

Frank Harold is an affiliate professor in the Department of Microbiology, University of Washington Health Sciences Center. Now retired, he remains engaged with science as a writer and unlicensed philosopher.

References

Harold FM. 2005. Molecules into cells: specifying spatial architecture. *Microbiol Mol Biol Rev.* **69:**544-64.

Karsenti E. 2008. Self-organization in cell biology: a brief history. *Nat Rev Mol Cell Biol.* **9**:255-62.

Lartigue C, Glass JI, Alperovich N, Pieper R, Parmar PP, Hutchison CA 3rd, Smith HO, Venter JC. 2007. Genome transplantation in bacteria: changing one species to another. *Science.* **317**:632-638.

Liu AP, & Fletcher DA. 2009. Opinion: Biology under construction: in vitro reconstitution of cellular function. *Nature Reviews Molecular Cell Biology* **10**:644-650 (September 2009).

Martin SG. 2009 Microtubule-dependent cell morphogenesis in the fission yeast. *Trends Cell Biol* **9**:447-454.

Ramamurthi KS, Lecuyer S, Stone HA, Losick R. 2009. Geometric cue for protein localization in a bacterium. *Science.* **6**:1354-1357.

Terenna CR, Makushok T, Velve-Casquillas G, Baigl D, Chen Y, Bornens M, Paoletti A, Piel M, Tran PT. 2008. Physical mechanisms redirecting cell polarity and cell shape in fission yeast. *Curr Biol.* **18**:1748-1753.

October 12, 2009

bit.ly/1LAnDEA

5

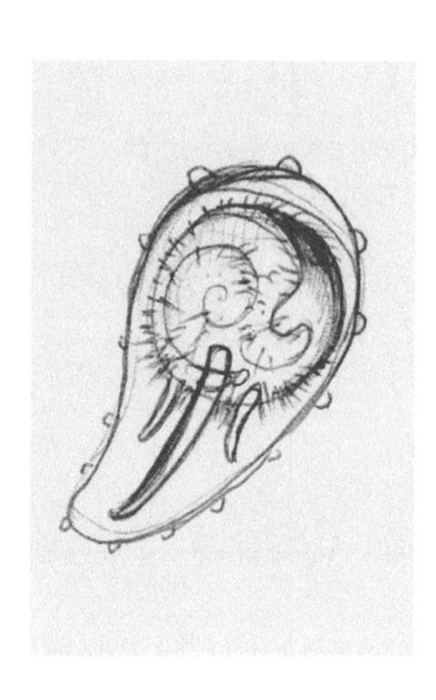

The Age of Imaging

by Elio

Not so long ago, it would have seemed implausible that biology would return to its origins as a visual science. Some would have considered this a regression to the days when biologists were pretty much confined to studying just what they could see, such as the shapes of organisms and their tissues. Back then, they focused on refining what Pliny had observed with his bare eyes, what Hooke and Leeuwenhoek saw under the microscope. The methodological lines of attack were dramatically redirected from the visual by the revolutionary discoveries of the second half of the last century. Biochemistry, genetics, molecular biology—none of them relied primarily on visualizing the structure of objects. For some time, doing morphology was suspect and, in some quarters, even using a microscope was equated with doing old-fashioned science.

How biology has (once again) changed!

Some of the most fundamental work done now once again involves seeing shapes and forms. Granted, genomics and its -omical kinfolk can be done with one's eyes closed (but with one's mind open). However, if you look no farther, you will miss much of the excitement of the day. Nowadays, mind-blowing insights come from seeing with your own eyes.

Biological imaging today starts with the very small, at the level of molecules—a field where splendid advances are being made. A new name, Structural Biology, was awarded to this sort of study.

In my graduate student years over half a century ago, only the rare visionary predicted that we would readily "see" how an enzyme works or how macromolecules interact with molecules large and small! These are grand achievements indeed. It gets better: single molecule imaging methods allow us to visualize the tiny movements made and the forces generated by proteins or ribosomes. One can now "see" in real time polymerases polymerizing and ribosomes translating.

Moving up a bit in magnitude, microscopy can also claim amazing developments. In my days, it was believed that the optical microscope had reached its physical limits and that the electron microscope had severe limitations. Recent progress on both these fronts continues at a stunning pace. Fluorescence techniques, including methods to clean up their signals, permit us to see single molecules in action at an exceptional degree of resolution, often in living cells. And the signals can even be quantitated. On the horizon are other techniques under development that hold promise for even greater resolution.

Newer on the scene is the coupling of cryotomography with the electron microscope, a technique that permits one to visualize the interior of unfixed whole cells. In a sense, this lets one crawl inside a relatively untreated cell, take a look around, and see what there is to see. I am reminded of an old prelim exam question that I had used to torment graduate students: "*If you could get to be small enough to fit inside a bacterium, what would you see*?" We thought this a "cool" question that paralleled the science fiction movie *Fantastic Voyage*, where a submarine with crew is miniaturized to 1 μm in length and thus able to travel the bloodstream of its inventor to destroy a blood clot in his brain. How about that! My question is no longer in the realm of science fiction! Although the technique doesn't miniaturize the experimenter, the result is the same: one can *pretend* to see what's inside a bacterium. The caveat in this statement is due to several factors: the cells have to be quite thin (although most prokaryotes in nature probably qualify); not all the structural constituents can be resolved with the same clarity; and the high voltage electron beam used probably introduces distortions. Still, crawling inside a bacterium is, by any standard, a magnificent achievement. So, what is there to see inside a "simple" bacterium? This will be the topic for a future posting.

The Age of Imaging is just beginning. It's hard to predict where it will lead, as the limits seem to constantly recede. Let's go for broke: someday we should be able to enjoy movies that show what goes on inside *living cells* at the resolution of the electron microscope. Maybe even talkies?

Elio is a Distinguished Professor Emeritus at Tufts University and an adjunct professor at San Diego State University and the University of California at San Diego.

March 31, 2008

bit.ly/1Gno2gE

#2

by Elio

Why have nitrogen-fixing bacterial endosymbionts of plants not evolved into organelles ("chlorochondria" or "azoplasts")?

December 1, 2006

bit.ly/1W2RlfK

6

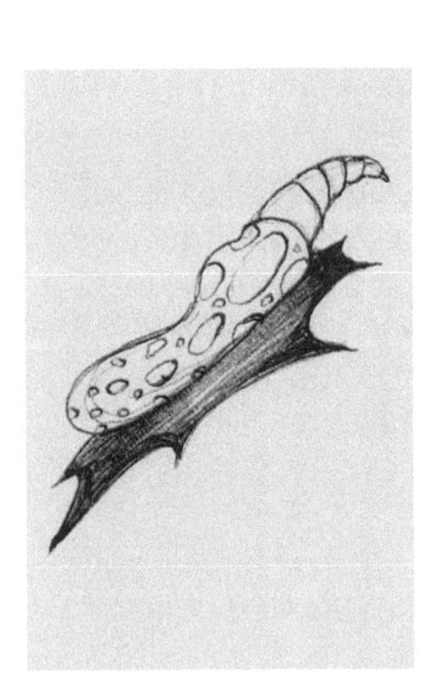

Bacillus subtilis: Wild and Tame

by Richard Losick

My dear friend Linc Sonenshein introduced me to *Bacillus subtilis* forty years ago when he was a graduate student with Salvador Luria. The remarkable capacity of *B. subtilis* to transform itself into a spore has been the focus of my research ever since. Before too long, Sonenshein and I focused on 168 and related strains, the *E. coli* K12 of the *B. subtilis* world. We did so for the reason that, thanks to the pioneering work of John Spizizen (with some magic from Charley Yanofsky and Norm Giles sprinkled in), strain 168 exhibited the remarkable capacity to take up DNA from its environment and recombine the DNA into its chromosome. This discovery of genetic competence opened the way to traditional and, eventually, molecular, genetics in *B. subtilis* and made the bacterium a premier model organism. At the same time, and what I did not realize until many years later, we also paid a price for using a strain that had been passaged many times in the laboratory.

Domestication has led to the production of long chains of sessile cells. Shown is a fluorescence micrograph taken by Dan Kearns of growing cells of laboratory *B. subtilis*. In addition to swimming cells (the green-colored singlets and doublets), the population contains many long chains of sessile cells. The cells were visualized with the vital membrane stain FM4-64 (red) and contained a fusion of the gene for the Green Fluorescence Protein (responsible for the green color) to a promoter under the control of a transcription factor that controls motility. Thus, only the motile cells in the image are green.

Wild (undomesticated) strains, in comparison, produce relatively few chains of sessile cells.

Ferdinand Cohn reported the discovery of *B. subtilis* in 1877. But the *B. subtilis* laboratory strains of today are a shadow of their former selves. Years and years of manipulation in the laboratory has robbed *B. subtilis* of much of its biology. On the one hand, laboratory strains can be transformed with DNA much more efficiently than undomesticated strains. On the other hand, laboratory strains are generally deficient in a variety of behaviors manifest in wild strains. Whereas wild strains are highly motile, have the capacity to swarm on surfaces, and form architecturally complex communities (biofilms), laboratory strains form long chains of sessile cells, fail to swarm, and form smooth colonies and thin pellicles. Indeed, studies with biofilms (in collaboration with Roberto Kolter) have changed our thinking about sporulation. We traditionally treated sporulation as largely a behavior of solitary cells, but recent work emphasizes the importance of studying spore formation in the context of multicellular communities (as has long been recognized in myxobacteria).

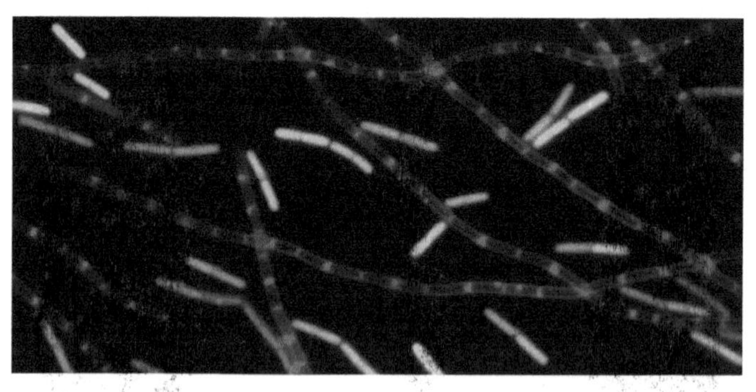

Domestication has led to the production of long chains of sessile cells. Shown is a fluorescence micrograph taken by Dan Kearns of growing cells of laboratory *B. subtilis*. In addition to swimming cells (the green-colored singlets and doublets), the population contains many long chains of sessile cells. The cells were visualized with the vital membrane stain FM4-64 (red) and contained a fusion of the gene for the Green Fluorescence Protein (responsible for the green color) to a promoter under the control of a transcription factor that controls motility. Thus, only the motile cells in the image are green. Wild (undomesticated) strains, in comparison, produce relatively few chains of sessile cells.

Source: **Kearns, DB, R Losick.** 2005. Cell population heterogeneity during growth of *Bacillus subtilis. Genes & Development* **19**:3083-3094.

How did this happen? I suspect that it is the result of heavy mutagenesis along with generations of researchers who, without thinking about it, favored rare smooth colonies that stayed put on the agar plate over other more unruly ones.

There are two lessons here that may be of general interest for those of us who consider small things. First, much biology may await discovery from revisiting the ancestral roots of popular laboratory strains. Second, this missing biology may hold the key to understanding the function of some of the myriad mysterious genes with which bacterial genomes are riddled. In short, back to the wild!

Richard Losick is Harvard College Professor, Maria Moors Cabot Professor of Biology at Harvard University.

September 15, 2008

bit.ly/1M2m15Q

#8

by Elio

Can you think of a place on Earth where there is free water but no microbes? (outside the bodies of other organisms or the lab)

March 2, 2007

bit.ly/1Gfpsdq

7

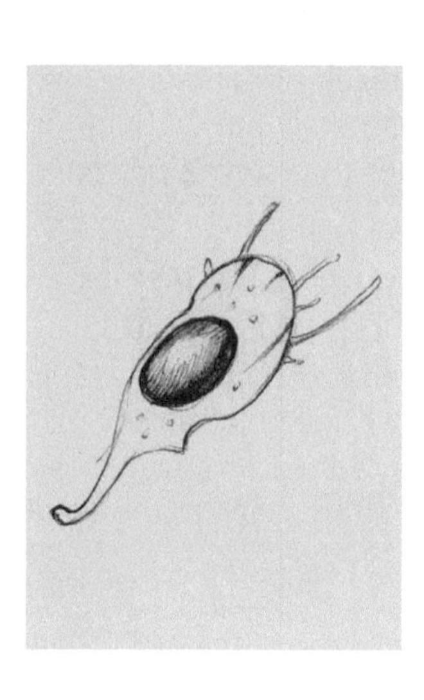

The Tyranny of Phylogeny: An Exhortation

by Elio

Two archaea walk into a bar and the bartender says, "If you guys are going to start in with the jokes again, Woese is me."

—Fred Rosenberg

There are days when I wish that the Woesian Three Domain scheme were wrong. Not that I would be happier if there were four or five or whatever number of domains. What would please me would be an escape from what I feel is an unnecessarily oppressive way of thinking, the seating of phylogeny (and its acolyte, genomics) alone at the head of the table. Why do I say this? Because as

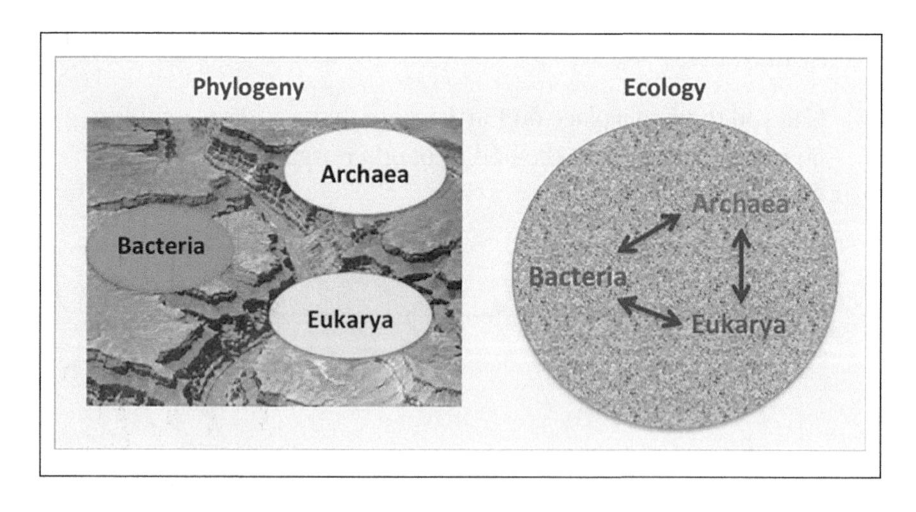

essential as phylogeny is to our understanding of the evolution of living organisms, equally vital is ecology to comprehend present day life. While it's good to know where you come from, it's equally important to know where you are and what you're doing there. The Spanish philosopher Ortega y Gasset said it well: "*Yo soy yo y mi circunstancia*" ("*I am I and my circumstance*").

A sole focus on phylogeny forces the past ahead of the present. Compare the two figures; they represent two different worldviews. One highlights the deep clefts between the three domains, the other is integrative and does away with such barriers. (The ecology perspective also acknowledges the highly abundant viruses, indicated by the stippled background.) Maybe it is not phylogeny's job to emphasize ecology, but neither should we be fixated on evolution alone. Obviously, no one is, so I apologize for erecting a straw man. Yet let me voice a wish: I would like to see a wider-ranging acknowledgement of each organism's give-and-take with its environment. Famed evolutionary biologist and geneticist Dobzhansky once said: "*Nothing in Biology makes sense except in the light of evolution.*" Do I dare modify this well-known dictum to read: "*Nothing in Biology makes sense except in the light of evolution and ecology.*"

October 25, 2012

bit.ly/1XgDLTc

8

Virus in the Room

by Welkin Johnson

I am the Lorax, I speak for the trees. I speak for the trees, for the trees have no tongues, and I'm asking you, sir, at the top of my lungs.

– from *The Lorax* by Dr. Seuss

As biologists, we divvy the biological realm up into domains using a formula that, frankly, smacks of nepotism, bestowing three glorious domains upon our closest relatives—the Eucaryota, the Archaea, and the Bacteria—while committing an injustice to the so-called viruses, lumping them together in a miscellaneous catch-all category ("viruses" from Latin for poison and other noxious substances) with contemptible disregard for phylogeny or any true measure of diversity.

Imagine that viruses, like Dr. Seuss's truffula trees, had a vocal advocate like the Lorax. Undoubtedly, through the agency of their outspoken mouthpiece, they would protest these gerrymandered borders and laugh at our skewed notions of biological diversity. After all (the viruses would argue), just consider the platypus, the coelacanth, the earthworm, and the bacillus. All these organisms have double-stranded DNA genomes, whose lengths all fall within roughly the same order of magnitude, which they replicate using evolutionarily customized versions of what amounts to the same basic enzymatic apparatus. How boring! How unimaginative! Now consider this (the viruses go on to say): the giant Mimivirus, 1256 nm of girth enfolding > 1,000,000 base pairs of DNA, and the tiny Circovirus, with a mere 1,800 bases of single-stranded DNA tucked inside a 20 nm-wide shell, are neither more nor less related to one another *than either one is to an elephant!* (For those who are not familiar with the elephant, it is a relative of the platypus, the coelacanth, the earthworm, and the bacillus.)

Let us thumb through the catalogue of viral genomes: here we find the familiar double-stranded DNA, including both linear and circular genomes, but also some with not-so-familiar twists—poxviruses, for example, covalently closing both ends of their linear double-stranded DNA genomes. We also find an abundance of themes not found anywhere among the domains of cellular life: thus, there are viruses with single-stranded DNA genomes and viruses with single-stranded RNA genomes, the latter including some that are negative-sense, some positive-sense, and some part positive and part negative (ambisense). Additionally, there are viruses with double-stranded RNA genomes, and if that isn't bizarre enough, there are viruses with segmented RNA genomes (to which the influenza virus belongs), whose virions incorporate a precise complement of eight different RNA segments.

Equally impressive are the reoviruses, with genomes composed of a dozen different segments of double-stranded RNA. Replicate that! And there are retroviruses, whose genomes are sometimes RNA (in the virion), and at other times double-stranded DNA (upon entering a host cell). Hepadnaviruses, possible cousins to the retroviruses, have gapped double-stranded DNA genomes with a bit of RNA thrown in, which they, too, convert to DNA by means of reverse transcriptase.

This diversity of genome styles each comes with its own uniquely-tailored replication system dictated in part by the need (shared by all viruses) to generate mRNA (because all viruses rely on host cells for translation). Importantly, there is very little if any phylogenetic evidence for a common ancestry connecting all the different viral types, or for grouping viruses together. Attempts to prove the existence of a last universal common ancestor of all viruses may be folly, as it is entirely possible that no such ancestor ever existed (that is, what we lump together as "viruses" actually represent uniquely evolved biological entities that happen, just by chance, to have taken on obligate intracellular parasitism as a mode of existence). At best, and by stretching the limits of phylogenetic comparisons, some of the RNA viruses can be combined into hypothetical "supergroups."

The tables thus turned, the viruses demand a fair redistricting, with the viral realm to include no fewer than seven domains to our three. They also ask that we wear name tags, since they are having trouble remembering how to tell an elephant from a bacillus.

Welkin Johnson is Professor and Chair of Biology at Boston College.

November 19, 2012

bit.ly/1GfyjMb

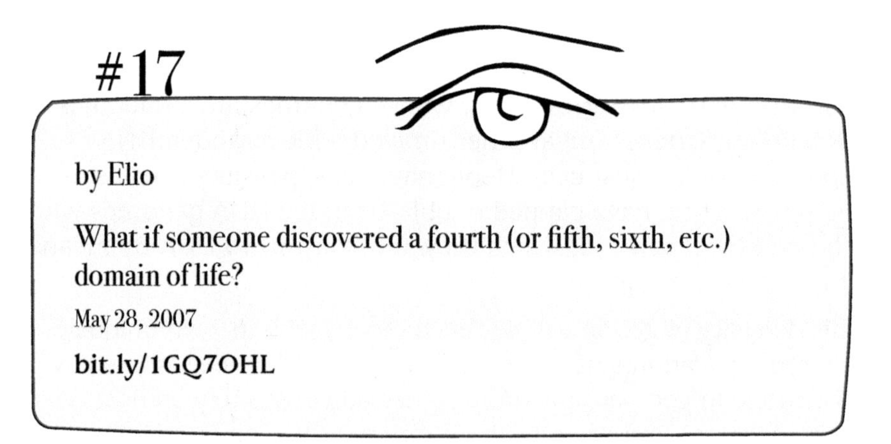

#17

by Elio

What if someone discovered a fourth (or fifth, sixth, etc.) domain of life?

May 28, 2007

bit.ly/1GQ7OHL

Feynman Said "Just Look at the Thing!"

by Jan Spitzer

On October 28, 2010, Elio posted this Talmudic Question: "*Richard Feynman, the famous physicist, said*: It is very easy to answer many of these fundamental biological questions; you just look at the thing! *To take him up on it, imagine a microscope that lets you observe single molecules in a living cell at one Angström resolution. What's the first thing you would do with it*?" Thank you, Elio, for allowing me to provide some thoughts on the matter from the perspective of a physical chemist/chemical engineer.

Such a microscope could indeed help address some of the fundamental issues in biology today. I must say up front that I am surprised that microbiologists would want to look at (small) molecules ("pure" chemistry), or at the chemical details of "bigger things," as suggested by the very notion of using a "Schaechter-Feynman supermicroscope." This hypothetical instrument would have a resolution of 0.1nm with exposure times in the picosecond range (making it a bit akin to an infrared spectrophotometer) and would operate in Feynman's quantum mechanical world. It would look at the chemistry of biology, dissecting cells into their molecular components that are then chemically characterized individually. But let me explain this more...

What Should We Look At?

Feynman suggested that "seeing better" is "better." However, what we see often depends on what we are looking for. At such very high resolutions, we risk focusing on details so small that we lose context and perspective, like looking at the leaves of individual trees and losing sight of the surrounding forest. We will also miss the lakes, the meadows, and even the blazing sunset—the reddish light scattering forward from the nucleating particles of the nano-fog, the beauty of which, Feynman insists, a mere poet may miss (1). Similarly, a narrowly focused molecular researcher overlooks the relationship of the detail to the living whole (2,3). I would suggest observing at a slightly larger scale, say dimensions of 10 to 100 nm and durations of a millisecond, to see the "metabolons," "modules," "hyperstructures," and the many functional protein complexes (signalsomes, stressosomes, transcriptomes, dividisomes, etc.) to find out if and how they exist as discrete physical objects. We might then see these large, transient biomacromolecular clusters appear and disappear, and see them interact at the subcellular level. We could take full advantage of the atomic resolution of this supermicroscope to visualize ionic currents of ATP, GTP, phosphates, K and Mg ions, bicarbonate, glutamate, etc., tracking them from their origins to where they sink and disappear. Even the vectorial ionic currents generated in the cell envelope with their movements through and around the large "omic" biomacromolecular clusters could be visualized as the cell grows. What would we see when a bacterial cell begins to die? Will the ionic currents "die"? (4-6). So many events to explore with our supermicroscope!

What Did Feynman Actually Say?

Feynman's suggestion betrays the physicist's weltanschauung, the world view that Feynman-influenced biologists focused on. At the time, however, neither biologists nor physicists understood what the "thing" to look at was. The situation then was similar to one that chemists had to face 100 years earlier (circa 1860)—having to develop methods of chemical analysis: first, qualitative—to know what *kind* of atoms and molecules one is dealing with, and then quantitative—how many, in what proportions, their structures, the chemical reactions between them, etc. It is worthwhile to reprint that section of Feynman's ("nanotechnology") talk:

> "*What are the most central and fundamental problems of biology today? They are questions like: What is the sequence of bases in the DNA? What happens when you have a mutation? How is the base order in the DNA connected to the order of amino acids in the protein? What is the structure of the RNA; is it single-chain or double-chain, and how is it related in its order of bases*

to the DNA? What is the organization of the microsomes? How are proteins synthesized? Where does the RNA go? How does it sit? Where do the proteins sit? Where do the amino acids go in? In photosynthesis, where is the chlorophyll; how is it arranged; where are the carotenoids involved in this thing? What is the system of the conversion of light into chemical energy? It is very easy to answer many of these fundamental biological questions; you just look at the thing! You will see the order of bases in the chain; you will see the structure of the microsome."

For the record, the "microsomes" contained ~20 nm "granules," which were eventually purified and characterized as today's ribosomes.

Artifacts, Artifacts...

Obviously, looking at the "thing" is not that simple. The "thing" could be an artifact of sample preparation. The history of electron microscopy has provided us with many such artifacts (7,8), and this remains an issue today for the new spectroscopic *in vivo* methods (9,10). There is also the sampling problem—the variability between and within cellular populations. It is challenging to maintain populations of "model" bacterial strains "frozen in evolution" and "constant," therefore reproducible between different laboratories. In addition, the physiological state of a cell population varies with the time of sampling, as well as with environmental and nutrient conditions (11,12). This would be a shock for Feynman, who thought about the "thing" as something static, picturing biomacromolecules as "sitting" somewhere. This is important for nanotechnology because when things "do not sit," "molecular hell breaks loose" thanks to the second law of thermodynamics—the entropy (disorder) seeking to maximize itself. Indeed the greatest achievements of nanotechnology have been so far mainly in the areas where molecules do "sit," i.e., in the solid state.

The Sum of the Parts

Which brings me back to the "reductionist's problem" encountered when attempting to reduce ("soft matter") biology to chemistry. Can we really understand a "living" system (a bacterial cell) by taking it apart, then identifying all its components by their chemistry and structure? This, after all, is what X-ray crystallography and NMR are achieving. The biochemical and physiological functions of many proteins and nucleic acids have been elucidated through enzymology and genetic studies (mutant analysis, for the most part). But to reiterate Craig Venter's observation: "*No single cellular system has all of its genes understood in terms of their biological roles*" (13). In fact, a protein may have two (or more) functions (moonlighting proteins); e.g., glutamate racemase, essential for providing the d-glutamate needed for bacterial wall synthesis, also inhibits DNA gyrase. I would argue that today's fundamental challenge in cellular biology is that RNA and proteins do not "sit"; they move, as does ATP, ions, water molecules, DNA, the ribosomes, even the cell itself. They all move on time scales that span about 10 orders of magnitude, from picoseconds to seconds! And they move about in a crowded environment, where molecules and macromolecules in very close molecular proximity interact with each other, breaking and making chemical bonds, non-covalently attracting and repelling one another with molecular forces ranging during the cell cycle from the ultra-weak (transient, non-heritable) to the very strong (permanent, 'heritable') (4,5,14-17).

The Humpty Dumpty Problem

The great achievement of the reductionist paradigm, of the taking apart of living things, is the demonstration that all we see there is just "ordinary" chemistry and physics. The chemistry of the lipids, proteins, and particularly the nucleic acids (the genetic code) is the 'same' for all living things; unique living systems are each a variation on the same chemical theme composed from relatively few chemicals—the "building blocks of life" (12). These biomolecules and biomacromolecules have been synthesized by non-biological methods of organic chemistry, starting with Wöhler's synthesis of urea in 1828 and culminating with the chemical synthesis (and enzymatic assembly) of the chromosome of the bacterium *Mycoplasma mycoides* (13). There can be little doubt that all such synthetic biomacromolecular products will have the same properties (chemical and biological) as those synthesized enzymatically by cells. However, it is when the molecules and macromolecules come together in a dynamic environment that the "living state" of matter is born. Because we do not yet know how the "living state" of matter comes about, we are astonished at seeing bacterial cells "do what they do"—growing and dividing at an incredible speed, sometimes in less than an hour! Even more astonishing is the fact that cellular populations of progenotes and universal ancestors (18,19) emerged 3.5 billion years ago!

In my opinion, efforts should focus on "putting Humpty Dumpty together again" (20), synthesizing "life" (21), or coaxing "life" to emerge from a "biotic soup" (22), mimicking the emergence of the chemistry of life from the "prebiotic soup" about 3.5 billion years ago (23-25). These "synthetic" approaches will have to take into account the physical chemistry underlying the *in vivo* state of living cells, i.e., the transient interactions of crowded biomacromolecules within them, and their cyclic non-equilibrium nature (14-17,20). Here Richard Feynman's other, much later sentiment comes to the rescue: "*What I cannot build (create), I cannot understand.*" It is clear that microbiologists are in a unique position to devise new protocols to study and even create life, without having to chemically synthesize anything, by taking bacterial cells apart, "killing them gently," and then trying to revive the "biotic soup." There are no fundamental reasons why such protocols would not work, and they would remove some of the mystery from Pasteur's dictum "*Omne vivum e vivo.*" We just need to figure out how to handle the spheroplasts, protoplasts, the nucleoid, ribosomes, and large protein clusters! Play with them; make them assemble into a living condition. Admittedly, this may not be as easy as just "looking at the thing" but may well turn out to be quite valuable.

Jan Spitzer is R&D manager of Mallard Creek Polymers in Charlotte, North Carolina.

References

1. http://www.goodreads.com/quotes/127643-poets-say-science-takes-away-from-the-beauty-of-the
2. A Letter to the Blog by Franklin M. Harold: Let's Put Organisms Back in the Picture!
3. **Harold FM.** 2005. Molecules into cells: specifying spatial architecture. *Microbiol Mol Biol Rev* **69:**544–564..
4. **Spitzer JJ, Poolman B.** 2005. Electrochemical structure of the crowded cytoplasm. *Trends Biochem Sci* **30:**536–541.
5. **Spitzer J.** 2011. From water and ions to crowded biomacromolecules: in vivo structuring of a prokaryotic cell. *Microbiol Mol Biol Rev* **75:**491–506.

6. **Davey HM.** 2011. Life, death, and in-between: meanings and methods in microbiology. *Appl Environ Microbiol* **77:**5571–5576.

7. **Heuser J.** 2002. Whatever happened to the 'microtrabecular concept'? *Biol Cell* **94:**561–596.

8. **Powell K.** 2005. "Porterplasm" and the microtrabecular lattice. *J Cell Biol* **170:**864–865.

9. **Margolin W.** 2012. The price of tags in protein localization studies. *J Bacteriol* **194:**6369–6371.

10. **Swulius MT, Jensen GJ.** 2012. The helical MreB cytoskeleton in E. coli MC1000/pLE7 is an artifact of the N-terminal YFP tag. *J Bacteriol* **194:**6369–6371.

11. **Mika JT, Krasnikov V, van den Bogaart G, de Haan F, Poolman B.** 2011. Evaluation of pulsed-FRAP and conventional-FRAP for determination of protein mobility in prokaryotic cells. *PLoS One* **6:**e25664.

12. **Neidhardt FC, Ingraham JL, Schaechter M.** 1990. *Physiology of the Bacterial Cell.* Sinauer Associates Publishers, Sunderland, MA.

13. **Gibson DG, Glass JI, Lartigue C, Noskov VN, Chuang RY, Algire MA, Benders GA, Montague MG, Ma L, Moodie MM, Merryman C, Vashee S, Krishnakumar R, Assad-Garcia N, Andrews-Pfannkoch C, Denisova EA, Young L, Qi ZQ, Segall-Shapiro TH, Calvey CH, Parmar PP, Hutchison CA III, Smith HO, Venter JC.** 2010. Creation of a bacterial cell controlled by a chemically synthesized genome. *Science* **329:**52–56.

14. **Zimmerman SB, Minton AP.** 1993. Macromolecular crowding: biochemical, biophysical, and physiological consequences. *Annu Rev Biophys Biomol Struct* **22:**27–65.

15. **Miklos AC, Li C, Sharaf NG, Pielak GJ.** 2010. Volume exclusion and soft interaction effects on protein stability under crowded conditions. *Biochemistry* **49:**6984–6991.

16. **Mika JT, Poolman B.** 2011. Macromolecule diffusion and confinement in prokaryotic cells. *Curr Opin Biotechnol* **22:**117–126.

17. **Wang Q, Zhuravleva A, Gierasch LM.** 2011. Exploring weak, transient protein--protein interactions in crowded in vivo environments by in-cell nuclear magnetic resonance spectroscopy. *Biochemistry* **50:**9225–9236.

18. **Woese C.** 1998. The universal ancestor. *Proc Natl Acad Sci USA* **95:**6854–6859.

19. **Fox GE.** 2010. Origin and evolution of the ribosome. *Cold Spring Harb Perspect Biol* **2:**a003483.

20. **Gierasch LM, Gershenson A.** 2009. Post-reductionist protein science, or putting Humpty Dumpty back together again. *Nat Chem Biol* **5:**774–777.

21. **Szostak JW, Bartel DP, Luisi PL.** 2001. Synthesizing life. *Nature* **409:**387–390.

22. http://abscicon2012.arc.nasa.gov/abstracts/abstract-detail/can-a-living-system-self-construct-from-a-biotic-soup/

23. **Bada JL, Lazcano A.** 2003. Perceptions of science. Prebiotic soup--revisiting the Miller experiment. *Science* **300:**745–746.

24. **Sullivan WT III, Baross JA (ed).** 2007. *Planets and Life*. Cambridge University Press, Cambridge, UK.

25. **Spitzer J, Poolman B.** 2009. The role of biomacromolecular crowding, ionic strength, and physicochemical gradients in the complexities of life's emergence. *Microbiol Mol Biol Rev* **73:**371–388.

March 11, 2013

bit.ly/1NRTgiJ

by Elio

Richard Feynman, the famous physicist, said: *It is very easy to answer many of these fundamental biological questions; you just look at the thing!* To take him up on it, imagine a microscope that lets you observe single molecules in a living cell at one Angström resolution. What's the first thing you would do with it?

October 28, 2010

bit.ly/1MAlGIa

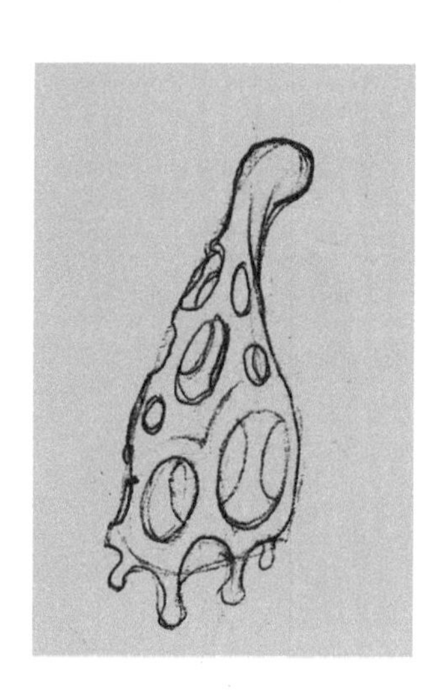

10

Self-Assembly for Me

by Elio

I have the grating feeling that the subject of self-assembly of complex biological structures may not always amass the level of respect it deserves. I reckon that its importance is generally appreciated but, as topics go, it tends at times to be set aside. Yet, this is one of the most magnificent aspects of biology, one that beautifully combines logic with mechanics and attests forcibly to the power of evolution. And it goes back a ways. The pioneering study on the self-assembly of phages played an integral role in the development of molecular biology.

Today, the assembly of the bacterial flagellar motor rates high on the list of exciting self-assembly phenomena, possibly vying with that of viral structure. The motor is a key constituent of bacterial flagella. It is located at the base of the structure and is responsible both for anchoring it to the bacterium and providing the mechanism for its rotation. It is a structure with many components, and its assembly constitutes an amazing engineering feat. One of the earliest indications of its complexity was recently exposed in these pages. Going back to 1971, purified flagella were convincingly shown to have an intricate base, consisting of several rings presumed to anchor the flagellum to the bacterial envelopes in a rotor-stator arrangement. This structural design for a molecular machine delightfully explained how flagella could both rotate and be kept in place.

Time passed since this spectacular early imagery, and with it came the development of techniques of previously unimaginable power.

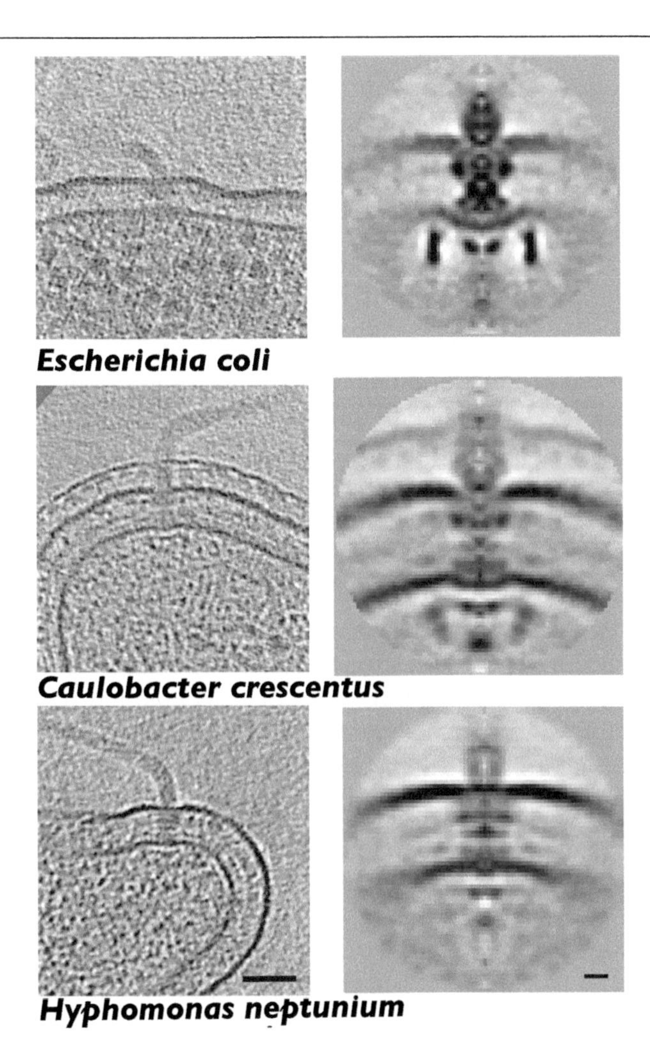

Flagellar motor structures obtained by electron cryotomography and subtomogram averaging. **Left column** 20-nm thick central slices through tomograms of individual cells exhibiting flagellar motors, arranged in the same order as they appear on a phylogenetic tree. Scale bar, 50 nm. **Right column** Axial slices through average reconstructions of each motor. Scale bar, 10 nm.

Source: **Chen S, Beeby M, Murphy GE, Leadbetter JR, Hendrixson DR, Briegel A, Li Z, Shi J, Tocheva EI, Müller A.** 2011. Structural diversity of bacterial flagellar motors. *The EMBO* **30**:2972-2981.

High among these is electron cryotomography, a way to reveal the 3D arrangements in unfixed biological material under the electron microscope. This is somewhat analogous to a CAT scan, but instead of the optical sections being parallel, they are produced by tilting the specimen at various angles. With this technique, along with the computerized analysis of single images (electron cryotomography and subtomography averaging), one can observe structures at 'macromolecular' (several nanometer) resolution. In other words, in exquisite molecular detail. The elements of the flagellar motor, the various rings, the center rod, the stator component, and what is known as the export apparatus, are now revealed in glorious detail. It's like looking at the wheel assembly of a car reduced about 10 million-fold.

Now comes a surprise. One would expect that such a complex structure be the product of an uncommon event in evolution, consequently, that it be alike in different bacterial species. Not so. A most exciting detailed analysis of eleven different species shows that although the basic plan is the same, these tiny machines vary considerably in detail. Their elements differ in curvature and in the positioning with regard to the axis. True, the bacteria species chosen included an assortment of flagellar arrangements, the flagella being polar in some, all over the surface (peritrichous) in others, and in yet others encased in the periplasm. One can well imagine that such different arrangements might require specially adapted machinery. But this finding does reveal a great degree of plasticity in the way flagellar motors are made. Isn't this amazing?

Self-assembly requires a high degree of "smartness" by the molecules involved—a higher degree than found in our "smartphones" that are all but self-assembled. Not only must the whole bunch of molecules carry out their intended function; they must be able to join with others into highly sophisticated ultra-tiny machines. Even more fascinating is that this self assembling ability is self-evolved! If I were starting over and wanted to dedicate myself to molecular mechanisms, I would be likely to turn to the study of such smart molecules.

August 27, 2014

bit.ly/1LHBIU5

11

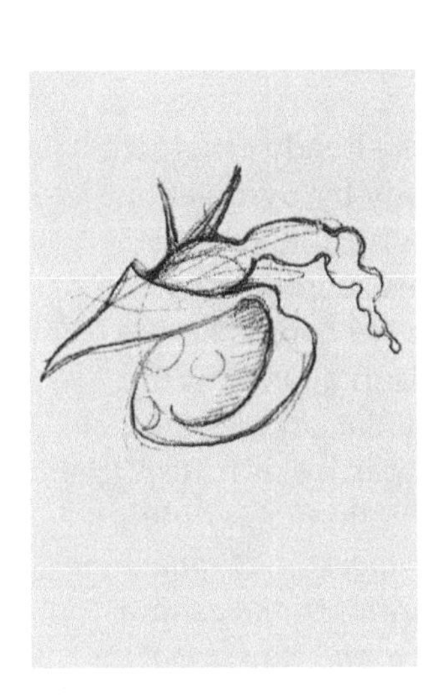

On the Definition of Prokaryotes

by Nanne Nanninga

As will be argued below, the present definition of a prokaryote is highly unsatisfactory. To give an example: a prokaryote is "*a cell or organism lacking a nucleus and other membrane-enclosed organelles, usually having its DNA in a single circular molecule*" (Brock, *Biology of Microorganisms*, 10th ed.). This seems a summary of the original definition of Stanier & van Niel (1962), which I quote for the sake of completeness: "*The principle distinguishing features of the procaryotic cell are: 1. absence of internal membranes which separate the resting nucleus from the cytoplasm, and isolate the enzymatic machinery of photosynthesis and of respiration in specific organelles; 2. nuclear division by fission, not by mitosis, a character possibly related to the presence of a single structure which carries all the genetic information of the cell; and 3. the presence of a cell wall which contains a specific mucopeptide as its strengthening element.*" Today's perception of these points amounts largely, as indicated above, to the absence of a nuclear envelope in prokaryotes. It should be mentioned that Stanier & van Niel in the above paper also wished to differentiate a bacterium from a virus and to incorporate blue-green algae within the prokaryotic domain.

The paper of Stanier & van Niel appeared more than 50 years ago. In this contribution I will attempt to present a more modern definition of a prokaryote, keeping in mind that one should distinguish between two aspects: its definition and its distinction from a eukaryote. But, before doing so I will insert an historic intermezzo.

Intermezzo

An alternative characterization of prokaryotes already emerged, though not explicitly, in 1964 through the study by Byrne et al. of the formation of DNA-ribosome complexes as analyzed by sucrose-gradient centrifugation. In those times sucrose gradient data were graphically expressed with radioactive counts on the vertical axis and fraction number on the horizontal axis. In the paper of Byrne et al. the focus was on the cellular organization of transcription and on the stability of RNAs, in particular mRNA (then also named cRNA). Their labeling and fractionation experiments led to the following conclusion: "*Our results suggest that some time after the onset of cRNA transcription on the surface of the gene, a ribosome will attach to the RNA molecule. This event may signal the beginning of polypeptide synthesis on the ribosome. The ribosome may then advance along the cRNA strand followed by other ribosomes, resulting in a continuous translation of the genetic message into polypeptide sequences.*" And further on: "*In a sense, ribosomes in bacterial systems may be a rudimentary form of the "carrier" ribonucleoprotein particles believed to transport cRNA from the nuclei of higher organisms.*" An electron micrograph of one of the fractions shows clusters of particles with ribosomal dimensions and threads, presumed to be DNA and perhaps mRNA.

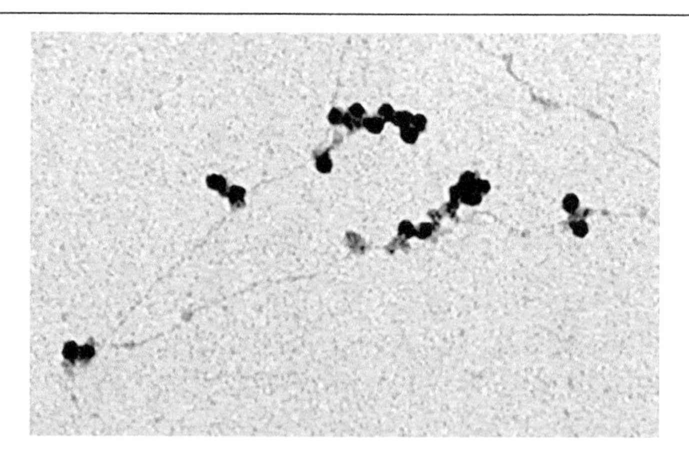

Dispersed strands of genomic DNA with polysomes attached of the hyper-thermophilic archaeon *Thermococcus kodakaraensis* visualized by the Miller technique (1970). Bar: 200 nm.

Source: **French SL, Santangelo TJ, Beyer AL, Reeve JN**. 2007. Transcription and translation are coupled in Archaea. *Molecular Biology and Evolution* **24**:893-895.

Protection of nascent transcribed RNA was also a concern of G. S. Stent (1966): "*What kind of system might assure* in vivo *removal of the nascent RNA from the template?*" In the case of mRNA, ribosomes came to the fore and mRNA-protection became incorporated into the framework of coupled transcription and translation; that is, translation starts before transcription has been finished. The conceptual drawing of Stent became reality through the impressive electron micrographs of Miller et al. (1970).

Then, *E. coli* was still the model organism. Some years ago, coupled transcription and translation was visualized for the hyperthermophilic archaeon *Thermococcus kodakaraensis* using the Miller technique (French et al. 2007). Thus, one can conclude that in the bacterium *E. coli* and in the archaeon *T. kodakaraensi* coupled transcription and translation are facts. Whether this applies to all Archaea remains to be seen. Because eukaryotes lack coupled transcription and translation, prokaryotes, in this way, distinguish themselves positively (Martin & Koonin, 2006).

The Distinction between Prokaryotes and Eukaryotes

As mentioned above, one should differentiate between definition and distinction. The fact that eukaryotes do not possess coupled transcription and translation—a negative qualification for some, by

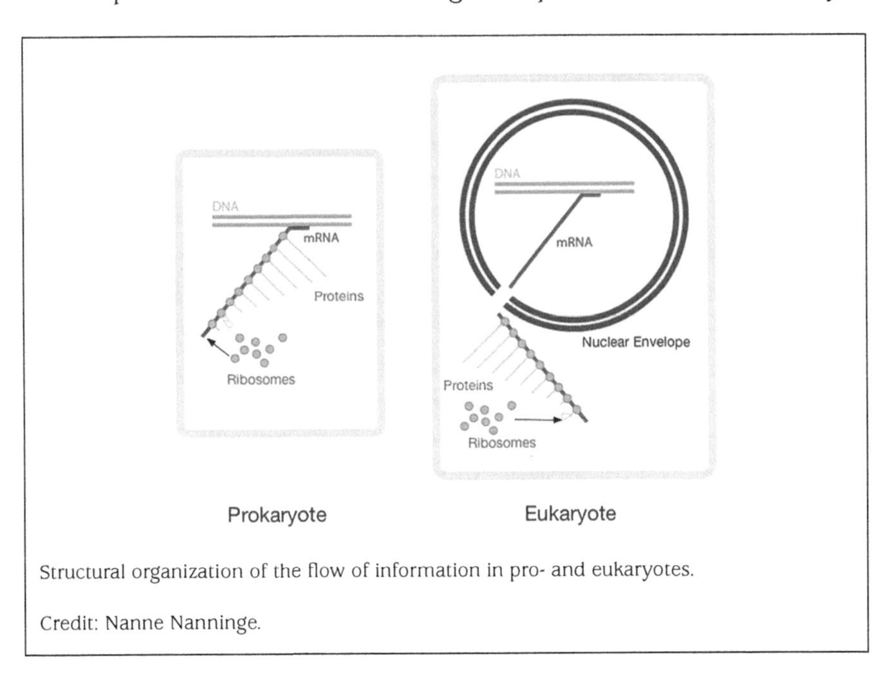

Structural organization of the flow of information in pro- and eukaryotes.

Credit: Nanne Nanninge.

the way—does not tell us much about eukaryotes. Also, the absence of a nuclear envelope in prokaryotes—a negative qualification, too—is not very informative. What is fundamental in cells relates to the structural organization of information processing, i.e., the flow of information from DNA through mRNA to protein. Conceptually, this bears on the central dogma of molecular biology. Thus, the main distinction between pro- and eukaryotes lies in the presence or absence of coupled transcription and translation, respectively. Does this suffice as a definition of a prokaryote? This question is easier posed than answered. I refer to another quotation from Stanier & van Niel in the same paper mentioned above: "*The differences between eucaryotic and procaryotic cells are not expressed in any gross features of cellular function; they reside rather* in differences with respect to the detailed organization of the cellular machinery" (italics by Stanier & van Niel). The absence of a cellular structure (a nuclear envelope) can hardly be considered an informative statement when it is known that transcription and translation are spatially coupled. The latter reflects the "*detailed organization of the cellular machinery*" and, I believe, this should be part of a definition, as should microscopic size and a cell wall. If one contrasts prokaryotic cytoplasm with that of a eukaryote, the presence of membrane-bounded organelles suspended in a dynamic cytoskeletal framework appears a dominant feature in the latter case. Such structural differentiation could hardly be accommodated for in a microscopic cell. Also remember that mitochondria and chloroplasts have prokaryotic dimensions. Following this reasoning, *a prokaryote can be considered a walled cell of microscopic size possessing coupled transcription and translation and a not-well differentiated cytoplasm*. According to our present knowledge, this applies to Bacteria and Archaea. Structurally, there is a "*deep gulf*" (Lane, 2011) between prokaryotes and eukaryotes. To what extent this bears on their phylogenetic relationships remains to be seen.

Two observations seem at variance with the foregoing. First, there is the remarkable membrane compartmentation of a planctomycete bacterium like *Gemmata obscuriglobus*. However, three-dimensional reconstructions suggest that the visual compartments are due to a sectioned lobed conformation of an *E. coli*-reminiscent organism (Santarella-Mellwig et al. 2013), implicating a prokaryotic organization. Second, nuclei do not seem to be devoid of protein synthesis, though the nature of the proteins involved are not known (David et al. 2012). Presumably, an exception that might prove the rule.

Nanne Nanninga is Emeritus Professor of Molecular Cytology, Swammerdam Institute for Life Sciences, University of Amsterdam, The Netherlands.

References

Madigan MT, Martinko JM, Parker J. 2003. Brock Biology of Microorganisms. 10th edition. Pearson Education, Inc., Upper Saddle River, NJ, USA.

Byrne R, Levin JG, Bladen HA, Nirenberg MW. 1964. The in vitro formation of a DNA-ribosome complex. *Proc Natl Acad Sci USA* **52:**140–148.

David A, Dolan BP, Hickman HD, Knowlton JJ, Clavarino G, Pierre P, Bennink JR, Yewdell JW. 2012. Nuclear translation visualized by ribosome-bound nascent chain puromycylation. *J Cell Biol* **197:**45–57.

French SL, Santangelo TJ, Beyer AL, Reeve JN. 2007. Transcription and translation are coupled in Archaea. *Mol Biol Evol* **24:**893–895.

Lane N. 2011. Energetics and genetics across the prokaryote-eukaryote divide. *Biol Direct* **6:**35.

Martin W, Koonin EV. 2006. A positive definition of prokaryotes. *Nature* **442:**868.

Miller OL Jr, Hamkalo BA, Thomas CA Jr. 1970. Visualization of bacterial genes in action. *Science* **169:**392–395.

Santarella-Mellwig R, Pruggnaller S, Roos N, Mattaj IW, Devos DP. 2013. Three-dimensional reconstruction of bacteria with a complex endomembrane system. *PLoS Biol* **11:**e1001565.

Stanier RY, Van Niel CB. 1962. The concept of a bacterium. *Arch Mikrobiol* **42:**17–35.

Stent GS. 1966. Genetic transcription. *Proc R Soc Lond B Biol Sci* **164:**181–197.

November 2, 2014

bit.ly/1RTn3q2

12

The Microbial Nature of Humans

by Maureen O'Malley

Quite a lot has been written about human microbiome research and how it changes older ideas about human autonomy, individuality, and identity (e.g., Brüssow 2015; Hutter et al. 2014; Pradeu 2014). Most of these discussions focus on how the biological basis for our "self" is in fact a consortium of different lineages of cells, and that the majority of these cells are microbial. Microbiome research has revealed how particular compositional patterns in the microbial gut communities of humans are associated with phenotypic characteristics, such as disease states. Experiments in mice have added weight to interpretations that many of these associations are causal, and that the gut microbiota is the cause of multiple human phenotypes. Although the direction of causality is seldom clear and probably goes both ways, the potential phenotypic effects of microbes on human characteristics have reinforced even more strongly the notion of human identity as deeply microbial. Indeed, some literature might make a reader think that all that matters for any human characteristic is the gut microbiota in and of itself.

In a recent commentary, Harald Brüssow (2015) has raised the issue of whether *"we are at a new level of defining human nature as a human-microbe consortium"* (p. 13). He suggests this question can only be answered when there is stronger evidence for cooperative causal effects on human phenotypes of mutualistic members of our microbial communities. In other words, mutualism is defined by selective effects and not by accidental benefits (West et al. 2007).

Only in such selected cooperative interactions, says Brüssow (p. 14), should we consider a microbial collective an "alter ego" of ourselves.

This is a well-reasoned conclusion, and one that takes microbiome reflections beyond descriptive statistics of the numbers and relative proportions of microbes in the human body. Useful as such descriptions are, they are limited in what they say about the biological relevance of the microbiota. Consideration of selected cooperative relationships produces explanations of why these interactions matter, how they persist, and what the system under consideration actually is. Presumably, large numbers of the microbes in the human body fall by the explanatory wayside as "mere" commensals and adventitious parasites—large in number but not forming a "unit" that can be explained evolutionarily.

Colorful shore of the Grand Prismatic Spring, Yellowstone National Park.

Credit: Frank Kovalchek.

https://www.flickr.com/photos/72213316@N00/3647773704

But does accepting an evolutionary explanation of some microbiota relationships really point to human "nature"? The idea of human nature is much disputed in philosophy, especially if it invokes anything like an essentialist natural kind (Kronfeldner et al. 2014). Kinds like this are not well regarded by micro- and other biologists, knowing what they do about biological variation and evolution. So Brüssow's evolutionary criterion might be proposing an alternative viable definition of what human nature is: something determined by selected functional relationships with microbes inhabiting human bodies.

Is this collective of human and microbiota a unit of selection in its own right? Unlike the integration of mitochondria (and later plastids) into what we now know as eukaryotic cells, the human body and its cooperative microbiota do not form a single unit of selection. The microbes retain reproductive autonomy, even if they rely on humans to provide a certain environment. Microbiota compositions are not strongly heritable either, although some of their functions may be. Humans reproduce separately from their microbes, and for the most part acquire them during and after birth. This separation of germ lines and lineages in time and space is usually the reason given for seeing humans and their microbiota–even the genuine mutualists—as separate Darwinian individuals or units of selection (Godfrey-Smith 2013).

Does this undermine the idea of humans as consortia of microbial and human cells? Only if being a Darwinian individual is the criterion for this designation. Instead, we might want to look at a spectrum of inevitable interactions with microbes, functional dependencies, and evolutionary persistent relationships. We know humans will not be germ-free except in the most extreme artificial situations. Some of these interactions will be beneficial and others harmful. Many of the most important interactions will not occur with microbes occupying the human gut or other bodily parts. Our place in the world comes with an obligation to interact with microbes at numerous levels, within and without our bodies. We depend on microbial by-products to survive, and we are nodes in larger webs of microbial interaction that define the planet. From this perspective, we are indeed "microbial"—inside, outside, and all around. Microbiome research has nudged our knowledge of ourselves so that we understand we occupy some peripheral node in this larger scheme, far from the center of biological interactions on the planet. Our node will eventually disappear, but the extinction of *Homo sapiens* will not lead to the collapse of the global ecosystem (whereas our persistence will continue to produce major problems for macroorganisms).

From this extended ecological point of view, we are embedded in a network of microbial interactions. This view is also reinforced from a phylogenetic perspective. If eukaryotes are indeed an odd kind of archaeon (Koonin and Yutin 2014; Spang et al. 2015) and humans just a very tiny twig in the eukaryote tree (Adl et al. 2012), then microbial phylogenetic knowledge allows us to realize even more conclusively our dependently evolved and maintained nature.

Microbiome research and how it draws our attention to the organisms within us should not blind us to broader microbial ecology and evolution. There is a great deal of non-human-focused microbiome research currently unraveling the major evolutionary and ecological ties we have to the microorganisms running the planet. This, I suggest, is where "human nature" studies are ultimately located, and where they could help make more sense of our own glorification of our gut microbiota.

Maureen O'Malley is a Senior Research Fellow in the Department of Philosophy, University of Sydney, and author of the book Philosophy of Microbiology.

References

Adl SM, Simpson AG, Lane CE, Lukeš J, Bass D, Bowser SS, Brown MW, Burki F, Dunthorn M, Hampl V, Heiss A, Hoppenrath M, Lara E, Le Gall L, Lynn DH, McManus H, Mitchell EA, Mozley-Stanridge SE, Parfrey LW, Pawlowski J, Rueckert S, Shadwick RS, Schoch CL, Smirnov A, Spiegel FW. 2012. The revised classification of eukaryotes. *J Eukaryot Microbiol* **59:**429–493.

Brüssow H. 2015. Microbiota and the human nature: know thyself. *Environ Microbiol* **17:**10–15.

Bouchard F, Huneman P (ed). 2013. Vienna Series in Theoretical Biology, p 17-36. *In From Groups to Individuals: Perspectives on Biological Associations and Emerging Individuality.* MIT Press, Cambridge, MA.

Hutter T, Gimbert C, Bouchard F, Lapointe FJ. 2015. Being human is a gut feeling. *Microbiome* **3:**9 10.1186/s401.68-015-0076-7.

Koonin EV, Yutin N. 2014. The dispersed archaeal eukaryome and the complex archaeal ancestor of eukaryotes. *Cold Spring Harb Perspect Biol* **6:**a016188 10.1101/cshperspect.a016188.

Kronfeldner M, et al. 2014. Recent work on human nature: beyond traditional essences. *Philos Compass* **9:**642–652.

Pradeu T. 2014. Galatea of the microbes. *Philosophers'. Magazine* **67:**89–95.

Spang A, Saw JH, Jørgensen SL, Zaremba-Niedzwiedzka K, Martijn J, Lind AE, van Eijk R, Schleper C, Guy L, Ettema TJ. 2015. Complex archaea that bridge the gap between prokaryotes and eukaryotes. *Nature* **521:**173–179.

West SA, Griffin AS, Gardner A. 2007. Social semantics: altruism, cooperation, mutualism, strong reciprocity and group selection. *J Evol Biol* **20:**415–432.

August 31, 2015

bit.ly/1OFxBuh

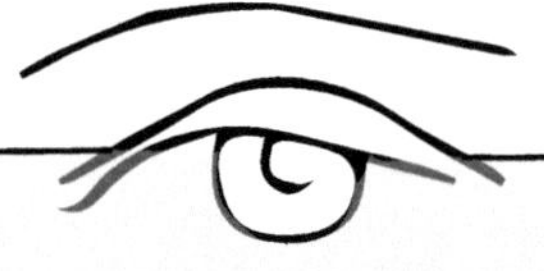

#6

by Elio

With the buzz created by the discovery of giant viruses, e.g., Mimivirus, the distinction between viruses and cells is said to get blurred. I maintain that this is not the case because the single most important distinction is that viruses lose their corporeal integrity, cells do not. What do you think? Are you a blurrer or a non-blurrer?

January 18, 2007

bit.ly/1Xgrzli

13

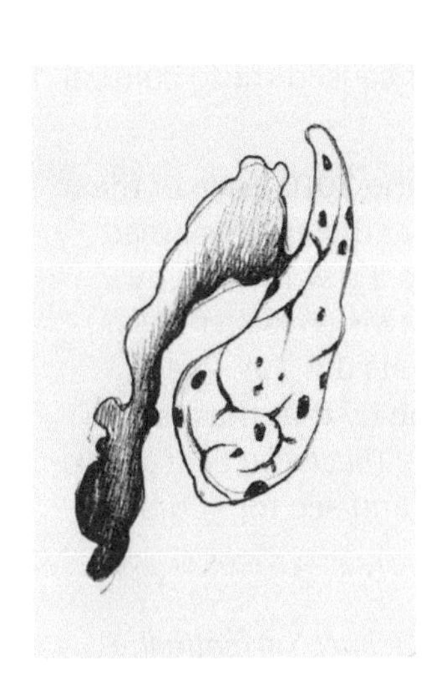

Good Writing Beats Bad Writing, Most Any Day

by Elio

There isn't any thought or idea that can't be expressed in a fairly simple declarative sentence...
– E.B. White, *Fro-Joy*

Just when some people believe that the world is going to hell in a handbasket, here I am, ready to make a cheerful personal statement: "*Scientific writing is improving!*" Of course I base this on reading the microbiological literature, but assume that it's generally true. What makes me say so? Well, more often than not, current research papers and reviews contain a fair share of simple declarative sentences. In my earlier days, typical statements were often in the passive form: "*The effect of X on Y has been studied.*" The first person form is now accepted, much to everyone's relief. And titles of articles tend to be informative. Gone is *Studies on the Metabolism of* Escherichia coli: *Part XIV*. And, although I couldn't swear to it, I believe that the language in graduate student papers has also improved. Nowadays, even humor is permissible. It has even permeated this blog, as exemplified by the wicked sense of humor of my co-blogger, Merry Youle, who has come up with titles such as *The Bacterium That Doesn't Know How to Tie Its Own Shoelaces, Coxiella Escapes from Cell,* and *A Holin One*.

What has happened? Have scientists turned into Thurbers overnight? Good question. I would guess that changes in culture (*sensu lato*) are the big factor. We seem to be more factual in communication and less prone to circumlocutions and Victorian adornments of speech. So, this may well be part of a general development, but what about its roots? What caused these cultural changes? I am no wiser.

This personal rapture was set off by perusing the 2010 issue of the *Annual Review of Microbiology*. It struck me that the articles I read were especially well written. I call as witnesses a few reviews by people I know. These include pieces by Kim Lewis (*Persister Cells*), Eric Rubin (and Michael C. Chao, *Letting Sleeping* dos *Lie: Does Dormancy Play a Role in Tuberculosis?*), and one by that master colorist of the microbial literature, Kevin Young (*Bacterial Shape: Two-Dimensional Questions and Possibilities*). Read and see for yourself.

Editors' Note:

We received the following response from Michael Yarmolinsky.

On the Quality of Scientific Writing

by Michael Yarmolinsky

Truth and clarity are complementary *

—Niels Bohr

Elio, in a posting of September 16, 2010, comments on the improved quality of scientific writing (in the literature of microbiology). Simple declarative statements are found in greater abundance; the passive voice is no longer used to avoid the first person; the titles of articles are more likely to come to the point and unlikely to be part of a numbered series; and humor is permitted. An examination of the implications and possible causes of the "changes in culture (*sensu lato*)" that Elio views with approval may interest, and perhaps surprise, the readers of this blog.

With respect to the declining use of the passive voice, one cause should be obvious. It is the insistent attention bestowed on sentences in the passive voice by grammar and spell check programs. Do not let a misguided word processor intimidate you. The impersonal style in expository writing is entirely appropriate. Your processor may underline "*bacteria were grown in LB*," but do not replace it with "*I grew the bacteria in LB*,"(Do indicate whether the recipe called for 5 g/l of NaCl or 10 g/l. It can make a difference.)

Elio cheerfully notes that titles have become more informative. The

* The irony of this clearly stated and often cited observation is that there appears to be some uncertainty as to whether Bohr has been correctly quoted.

abruptness and magnitude of the change is astonishing. A shift in the style of titles, from the indicative (denoting the subject) to the declarative (asserting that something is so) was documented in 1990 by J L Rosner.[1] Of 2582 titles published in the *Proceedings of the National Academy of Sciences* from 1960 to 1969, not one is a declarative title, but by 1980 25% were declarative and after 1986 the frequency had climbed to over 45% (2).

Yes, titles are more informative. But what about the reliability of the information provided? Therein lies the rub. Rosner entitled his article: *Reflections of Science as a Product*. As the title suggests, he saw the striking historical trend to reflect arrogance and even the loss of a guiding vision of science.

A tendency to view the science of biology as a product rather than a process has been bemoaned by various authors, most passionately by Carl R. Woese in an important essay: *A New Biology for a New Century* (2004) (3). Both Rosner and Woese place considerable blame on Jim Watson and his disciples, for whom science would seem to be necessarily goal-oriented. This belief appears to have taken hold because goal-oriented science has been extravagantly successful.

The issues raised by Rosner's article have been revisited more recently in medical journals, e.g., by Neville Goodman (2000) (4) and Jeff Aronson (2009) (5). Goodman noted an exponential increase in active verbs in the titles of clinical trial reports. He undertook a review of 24 reports of clinical interventions published in 1996 with "prevents" in their titles. In at least eight of these cases, the intervention had not been preventive at all. What makes this frightening is evidence, cited by Goodman, that physicians sometimes make clinical decisions from the titles of journal articles. I suspect that authors of reviews in which the bibliography runs to many hundreds of items occasionally take similar shortcuts. Goodman concludes: "*There may be arguments for reviews and editorials carrying informative [in the sense of declarative] titles, but they are too often wrong to have any place in the reporting of research. Journals should ask for indicative titles, or alter investigators' informative titles during sub-editing.*"

I find it comforting to know that the *Journal of Bacteriology* in its current *Instructions to Authors* disallows titles in numbered series and explicitly states: "*Avoid the main title/subtitle arrangement,* ***complete sentences,*** *and unnecessary articles*" (emphasis added). To my knowledge none of the other journals with which I am familiar recommend that their potential authors avoid declarative titles, although the rarity of such titles a mere forty years ago suggests that editorial pressure to stay with indicative titles may well have prevailed at the time.

Aronson notes with dismay that: "*in the last 15 years, declarative titles have found a new sponsor—evidence-based medicine. Using declarative titles was one of the declared procedures of the journal* Evidence-Based Medicine, *when it was launched in 1995, and in 2003,* The Journal of Clinical Epidemiology *started not only to publish declarative titles but also to commission them, claiming that such titles would be more informative. "Indicative' titles, they said, "give the purpose of a study," and "declarative" (or "informative") titles "give the conclusion." Although, as one declarative title has put it, "The evidence provided by a single trial is less reliable than its statistical analysis suggests"* (5).

Aronson, considering all that can be hidden from view in a declarative title, concludes: "*At best, declarative titles mislead; at worst they may enshrine a falsehood as a permanent truth.*" This may be overstating the case, but in my view, not by much. When editors succumb to the lure of the sound-bite mentality, it is up to authors to resist. There remains the topic of humor. It would be foolish of me to inveigh against humor, as I have ventured, more than once, to be witty in a scientific context. Humor can serve as an enticement to read on. However, much of what passes for humor in the titles or texts of articles strikes me as an embarrassment and should have been suppressed by an editor or reviewer or, better still, the author.

In finding the proper balance between truth and clarity, teacher and researcher are prone to make different choices. Elio has been and continues to be a superlative teacher; his blog is admired as a teaching aid. My career in science has been almost entirely in research. If my remarks reflect a rather jaundiced view of progress, so be it. I hope they will stimulate useful discussion.

References

1. **Rosner JL.** 1990. Reflections of science as a product. *Nature*, **345**:108.
2. A specific marker of the change is the near disappearance from contemporary literature of the "On" that graced the indicative titles of seminal works of previous generations: *On Growth and Form* (1916) by D'Arcy Thompson, *On the Origin of Species* (1859) by Charles Darwin, and all the way back more than two millennia to *De Rerum Natura* [*On the Nature of Things*] by Lucretius to name only three. I miss "On.....

3. **Woese CR.** 2004. A New Biology for a New Century. *Microbiol Molec Biol Revs* **68**: 173-186.
4. **Goodman, NW.** 2000. Survey of active verbs in the titles of clinical trial reports. *BMJ* **320**:914-915.
5. **Aronson, J.** 2009. When I use a word...Declarative titles. *QJM*; **103**:207-9. Available online

Elio Replies

Michael Yarmolinsky brings his powerful powers of insight to examine the reasons I mentioned for believing that the quality of scientific writing may be on the rise. He suggests that the greater use of the active form may be due to grammar checking programs. I wouldn't know as mine is turned off. Incidentally, I distrust the spelling program because it can't tell "does not" from "doe snot." Then come the titles. Michael agrees that they have become more informative but wonders if they may also have become less reliable. Guess what: there is even some data to suggest that. However, the studies cited by Michael are mainly of medical journals, some big on clinical trials. I don't really know, but I wonder if this applies to non-clinical science journals. Not to be flip about it, but despite some overlap the two have different aims, thus may respond to different societal pressures. The need for the papers to be convincing may be greater in the former than the latter. More "basic" articles don't need to "sell" themselves.

September 16, 2010

bit.ly/1LHBRXG

14

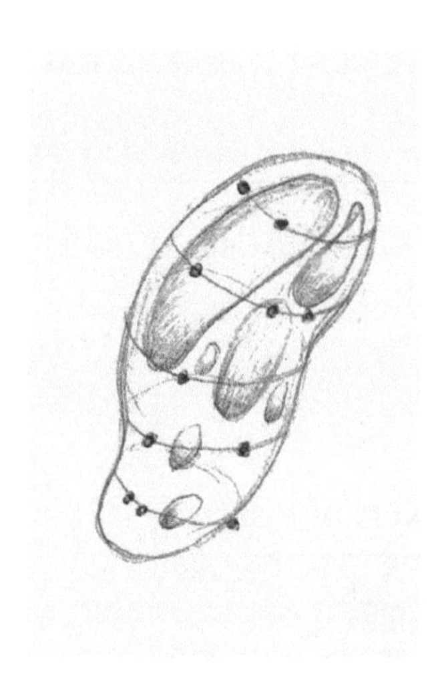

Finally, Farewell to "Stamp Collecting"...

by Christoph Weigel

A perspective paper by Margaret McFall-Ngai and colleagues was recently featured by Elio in this blog, strongly emphasizing its Chicxulub-like impact on microbiology. Here I offer a postscript, a few loosely connected thoughts from a historical perspective about its impact on biology and life sciences in general.

Until the '50's of the last century, advancement in biology was largely the product of three overlapping generations—students, active scientists, and emeriti—laboring over methods, paradigms, concepts, and theories. With few exceptions, these were European and North American men. Theories put forward by the emeriti during their active time tended to be overthrown by their former students who now become active scientists themselves: a spiral of slow progress. Since experiments were tedious and methodological progress slow, scientists were inclined to heated debates regarding concepts and theories. Few theories held for more than one generation, notable exceptions being Darwin's insight of evolution, Mendel's concept of inheritance, and the cell theory by Schleiden and Schwann. Collecting thousands of different mosses or pinning thousands of insects for a museum collection was considered at least equally important as experiments, the lattermost often designed to prove an existing theory rather than to generate a new testable hypothesis. Nevertheless, Louis Pasteur's experiments disproving the spontaneous generation of life and Robert Koch's postulates for proving disease causation can be considered to have ushered in the dawn of experimental biology.

By the late twentieth century, biologists were diligently striving to disprove Ernest Rutherford's famous dictum that "*science is either physics or stamp collecting.*" As molecular biology emerged, it was hoped—and rigorously asserted by its practitioners—that living systems could be completely understood in terms of the properties of their constituent parts (fundamentalist reductionism), and biology ultimately would be reduced to physics, as according to James Watson: "*... there are only atoms. Everything else is merely social work.*" Evolution research was dismissed because, being in essence historical, it could not be reduced to physics. Instead, the fitting approach was to solve the structures of biomolecules, which would thereby reveal their function. Likewise, the road to increased understanding of biology was the falsifiable working hypothesis, itself derived from previous experimental results (empirical reductionism).

Influenced to some extent by the *New Age* idea that our planet is most fittingly perceived *in toto* as a single living organism, a growing number of biologists in the '80s began to argue for a holistic approach. The reductionist approach then in vogue could not explain the emergent properties of complex biological systems, or, as Steven Rose phrased it: "*...watch a flock of birds, startled by a noise, take off from the field on which they have settled—see them wheel and turn in formation, and try to explain or predict the behaviour of the group merely from a knowledge of the wing-musculature of each individual and aerodynamic theory.*" However, there was a flaw in this argument: no holistic concept at that time was able to propose meaningful experiments.

At about the same time, a second criticism was put forward by biologists concerned that reducing biology to physics could in the end strangle scientific creativity. They favored curiosity-driven research over technology-driven research. Or, as Elio put it during a meeting in '87 in memory of Luigi Gorini: "*On the planet Krypton every experiment works. As a consequence people quickly run out of ideas and so they spend their time sequencing the human genome. With Luigi, experiments did not always work, but he never ran out of ideas.*"

Now, early in the twenty-first century, the situation is dramatically different. Never before were the life sciences explored by so many researchers from diverse cultural backgrounds, both men and women. Due to the ever-increasing speed of technological development, research now spans about five rather than three contemporary generations. Cutting-edge technology of the '80s is at best of historical interest today—who remembers Maxam-Gilbert sequencing? Experiments have become less tedious, but they now produce so much data for analysis that hardly any time is left for any of these multiple generations to debate what it all means. When the human genome sequence was published, biology hit a wall of biological complexity. Many biologists saw that fundamentalist reductionism was failing and the spiral of progress arrested as biology was drawn in various directions. Its central narrative seemed lost—almost.

At this point, enter the paper by McFall-Ngai et al., just in time, adding umami flavor to Carl Woese's call for "New Biology for a New Century." That call was paraphrased elegantly by Freeman Dyson: "*... postulating a golden age of pre-Darwinian life, when horizontal gene transfer was universal and separate species did not yet exist. Life was then a community of cells of various kinds, sharing their genetic information so that clever chemical tricks and catalytic processes invented by one creature could be inherited by all of them. Evolution was a communal affair, the whole community advancing in metabolic and reproductive efficiency as the genes of the most efficient cells were shared. ... But then, one evil day, a cell resembling a primitive bacterium happened to find itself one jump ahead of its neighbors in efficiency. That cell separated itself from the community and refused to share. Its offspring became the first species of*

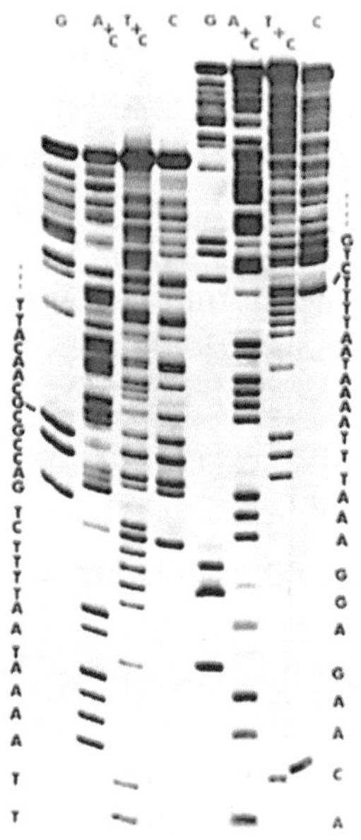

The sequence of the *E. coli rrnB* gene determined in 1979 by the Maxam-Gilbert technique.

bacteria—and the first species of any kind—reserving their intellectual property for their own private use. With their superior efficiency, the bacteria continued to prosper and to evolve separately, while the rest of the community continued its communal life." Although Margaret McFall-Ngai and her co-workers refrain from expressing it explicitly, I can easily imagine them adding to this narrative: ...*In separating itself from the community, refusing to share everything, this first species did not end communication with its siblings and the rest of the bunch, but rather increased its specificity, as witnessed by the ubiquitous communication among and direct interactions—even gene swapping—between the extant prokaryotes and eukaryotes, viruses, and a plethora of mobile genetic elements.*

To come full circle—or more precisely to reenter the spiral—Karl Popper had suggested already in 1986 that by adopting "active Darwinism" biology would avoid the teleological trap and eventually come into accordance with his scientific method of reductionism. Instead of the prevailing view in which selection was the imposed driving force of evolution, Popper's "active Darwinism" proposed that: "*...the organism itself is not passive and neutral, waiting to be selected, but instead actively participates in its own selection, by choosing appropriate environments and modifying inappropriate ones; organism and environment interpenetrate and modify one another in ways which are determined in part by their own mutual history.*" I assume Margaret McFall-Ngai and her colleagues would prefer the plural here: *organism**s** and environment interpenetrate and modify one another...* This added complexity can then be tackled by the approved methodologies of empirical reductionism without the danger of reverting to "stamp collecting,," as pointed out by Carl Woese.

This is where we stand today. Biology has its twenty-first-century narrative, which is just another word for an extended to-do list for biologists. The good news (for Elio, in particular): acute observation and curiosity have regained their pivotal role in finding out what life is all about. As we move forward, we eventually can teach computers one of the most precious, though enigmatic, of human traits—pattern recognition—so they can help us to cope with the approaching tsunami of data, help us visualize biological complexity.

Christoph Weigel is a lecturer in Life Science Engineering at the Hochschule für Technik und Wirtschaft, *Berlin, Germany, and an Associate Blogger.*

References:

McFall-Ngai M, Hadfield MG, Bosch TC, Carey HV, Domazet-Lošo T, Douglas AE, Dubilier N, Eberl G, Fukami T, Gilbert SF, Hentschel U, King N, Kjelleberg S, Knoll AH, Kremer N, Mazmanian SK, Metcalf JL, Nealson K, Pierce NE, Rawls JF, Reid A, Ruby EG, Rumpho M, Sanders JG, Tautz D, Wernegreen JJ. 2013. Animals in a bacterial world, a new imperative for the life sciences. *Proc Natl Acad Sci USA* **110:**3229–3236.

Rose S. 1988. Reflections on reductionism. *Trends Biochem Sci* **13:**160–162.

Woese C. 2004. A New Biology for a New Century. *Microbiol Mol Biol Rev* **68:**173–186.

June 10, 2013

bit.ly/1RYA8SO

#115

by Elio

Is global warming likely to result in a significant net increase, decrease, or no substantial change in the microbial biomass on Earth?

December 3, 2014

bit.ly/1hOOeoO

PART 2

Accounts of the Past

As science advances, stellar research of the day fades from the spotlight. Here I retrieve a few examples from the good old days.

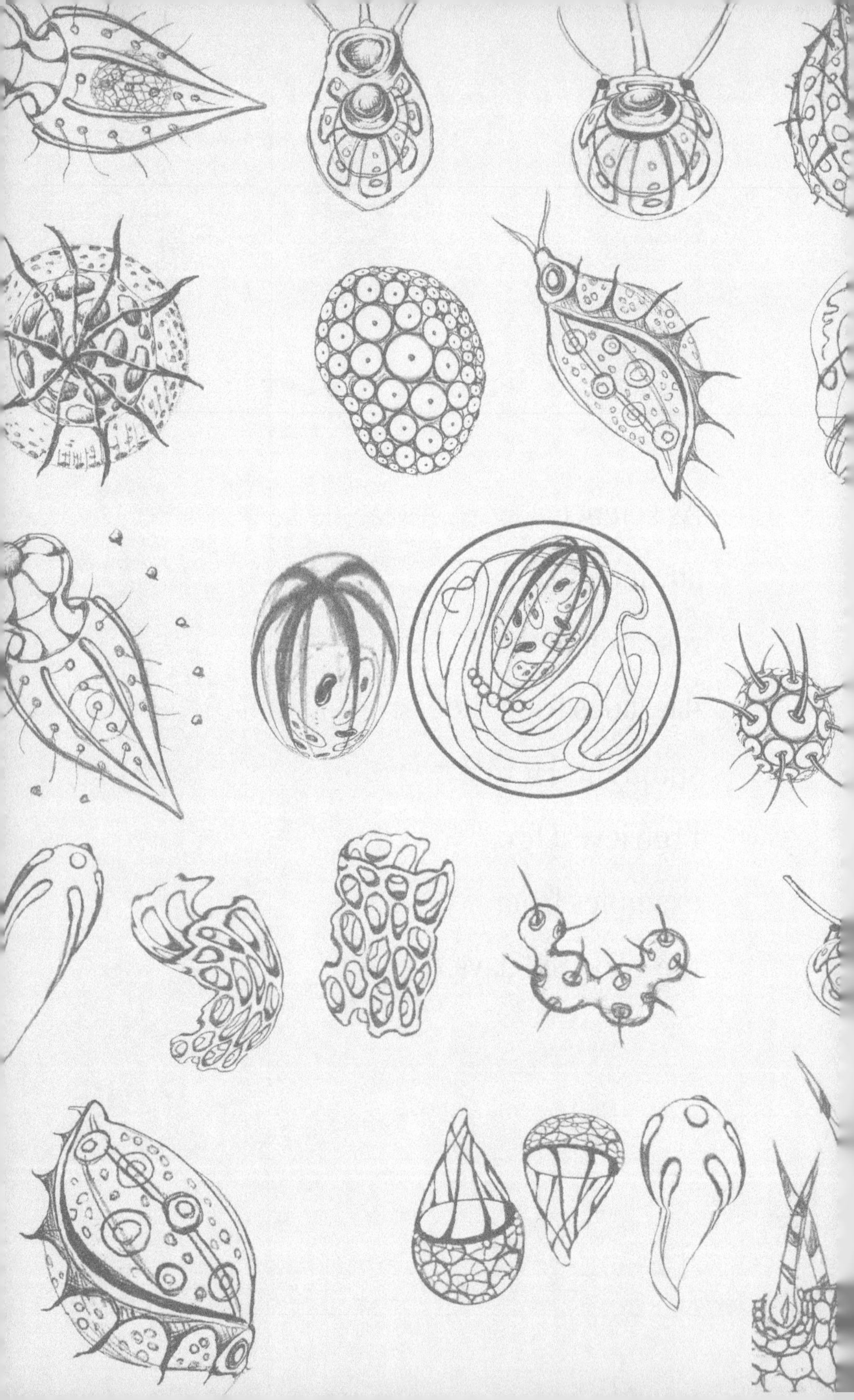

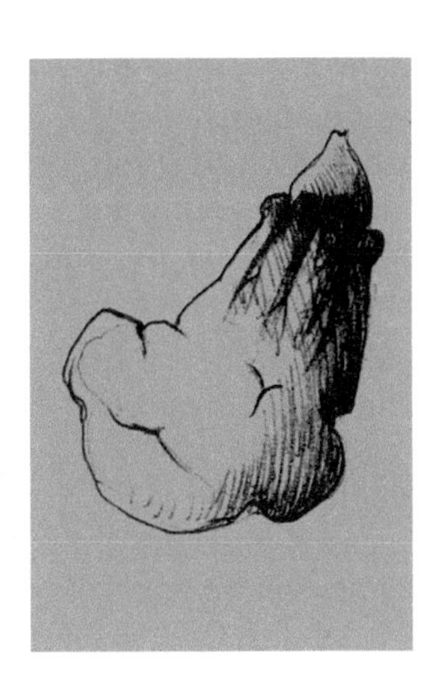

15

Esther Lederberg: Pioneer of Bacterial Genetics

by Mercé Piqueras

"*She did pioneering work in genetics, but it was her husband who won a Nobel prize.*" So said an obituary in the British newspaper *The Guardian* regarding Esther Lederberg, a North American microbiologist married to Joshua Lederberg from 1946 to 1966 (8). Being married to and working alongside such a phenomenal scientist was not only greatly stimulating; it was also a major handicap. As long as she was *Mrs.* Lederberg, the success of their joint work was attributed to her husband. At 31, Joshua was already a full professor whereas Esther, who was three years older, remained an associate investigator. In an interview that followed his 1953 Eli Lilly award in Bacteriology, Joshua affirmed that this prize should have been shared with his wife. Despite that, on the occasion of his Nobel Prize, Joshua did not mentioned Esther in either of his speeches, when he received the award and during the gala dinner.

Esther Miriam Zimmer was born in 1922 in The Bronx, New York. She attended Hunter College, intending to major in French or Literature. She changed her mind and opted for Biochemistry, ignoring the advice of her teachers who told her that science offered women few career opportunities. After finishing her studies, she worked for some time at the Carnegie Institution of Washington before going to Stanford University for a master's degree in genetics, working as a teaching assistant to be able to pay for her studies. Years later she told that at that time she had so little money that she ate the legs of the frogs used for dissection in a lab course.

In 1946, she finished her master's degree and, in December of that year, married Joshua. Although three years younger, he had gotten a professorship at the University of Wisconsin in Madison. Esther followed him there and worked with him while getting her doctorate in 1950. That year she discovered phage lambda, one the milestones in her research career.

In 1958 Joshua received the Nobel Prize and the same year moved to Stanford, where he founded and directed the Department of Genetics. Again, Esther followed her husband. For the rest of her professional career, Esther remained at Stanford, where she established the Plasmid Reference Center, which she directed until 1985, one year after officially retiring. Esther and Joshua got divorced in 1966.

Her love for medieval, Renaissance, and Baroque music led Esther to help organize an amateur orchestra in which she played the recorder. She kept up this interest to the end of her days, going to rehearsals and concerts even when her mobility problems forced her to use a walker. In 1989 he met Matthew Simon, who

Esther Lederberg in her home in Wisconsin (1958, the day that the Nobel Prize of Joshua Lederberg was announced).

shared her passion for music, and they were married in 1993. On November 11, 2006, Esther died of pneumonia.

Among her many achievements are discovering phage lambda, inventing the replica plating technique, and, along with Luigi Cavalli-Sforza from Milan, discovering the F plasmid (fertility factor or F factor). Nevertheless, the article that described the F plasmid has Joshua Lederberg as first author (4).

Phage Lambda

Esther discovered lambda, one of the best known coliphages, in 1950, the same year that she got her Ph.D. (1, 2). Lambda has a property that was not known for other viruses up to then: instead of multiplying rapidly inside the host cell and killing it, it integrates its DNA into that of the infected bacterium. Its genome is transmitted from one generation to the next without harming the host organism. In fact, what is transmitted is not the virus but the instructions to manufacture it. In certain circumstances, for example when the bacterium is stressed due to nutrient deprivation, the DNA of the

Esther Lederberg in her laboratory at Stanford.

virus is activated. The cellular machinery of the bacterium now begins to manufacture progeny viruses, thus killing the host cell. Thus, lambda has a dual behavior. Its two life styles are called lysogenic (when it maintains a stable genetic relationship with the host cell) and lytic (when it reproduces and causes lysis of the host cell). Lysogenic viruses are called temperate.

Esther discovered lambda by observing colonies of mixed cultures of *E. coli* K-12 and a strain that had been irradiated with ultraviolet light called W-518. Some colonies looked irregular, as if some segments were missing. This was due to the presence of the virus which, being latent in *E. coli* K-12, passed unnoticed while causing lysis in the mutant W-518 strain. At first it was believed that lambda was located in the cytoplasm following the model of the kappa particles of paramecium (it was later discovered that these particles are endosymbiont bacteria (7). But then Esther found that was incorrect; the data obtained suggested that lambda should be located on the chromosome, at a specific locus for lysogeny (Lp) linked to a Gal marker (Gal4). In an autobiographical account, Esther wrote that at first she worked on this project alone, but was helped by comments from Joshua and other members of the lab (10).

The ease with which lambda could be grown in *E. coli*, the fact that it was not pathogenic (except for bacteria), and the great amount of knowledge already available about its host soon helped turn it into a model to study other viruses that behave the same way, such as the herpes virus. In people suffering from cold sores, after a primary infection the virus is integrated into the DNA of cells in their body. In stressful situations, or when body defenses are low, the virus reveals itself again and herpes develops. The work of Esther with lambda elucidated phenomena such as the transfer of genetic material between bacteria (horizontal transfer), transduction (generalized or specialized), and many mechanisms of gene regulation. Lambda has also found multiple application in molecular biology. Among others, it is a marker of molecular weights, and can be used in building cloning vectors, and in engineering recombination (recombineering) (6). Also, in recent years phage therapy has become popular (it is not new, however; it was developed in the 1920's). With increasing resistance to antibiotics and the difficulty to find ones that replace those that lose effectiveness, phages are again in the crosshairs of antimicrobial research (8). Being among the best-studied phages, T coliphages and lambda serve as crucial models.

Replica Plating

Unlike other fields of biology, in microbiology the unit of study is usually not an isolated individual but a population—a colony on a Petri dish. Sometimes it is desirable to have identical bacterial cultures to compare their reactions to environmental changes such as in nutrition and temperature. But how to obtain sets of culture plates where the colonies of the same species are located on the same geometric location in different plates? It's like having a copy of a drawing or someone's signature using an inked rubber stamp. Several researchers had in fact used blotting paper and metal brushes with small prongs. In 1946, Joshua Lederberg learned from E. Tatum how to replicate colonies using toothpicks.

However, it was Esther who came up with a simpler solution. She thought to make identical replicas of a Petri dish by copying the original using velvet cotton pads. When pressing the original plate on the velvet, the little surface fibers of the fabric acted as tiny inoculating needles, transferring bacteria from each colony to the same position as in the plate from which they came. Esther checked different velvet cloths in a fabric store and chose the one whose thickness and "hair" were the most suitable for her purpose. She also tried different detergents to find the best one for washing them. The article that describes this simple yet innovative technique was published in 1952 and signed by Joshua and Esther Lederberg, with him as the first author (3). Relevant to the time, the replica plating method permitted isolation of antibiotic resistant mutants that had not been in previous contact with the drug, which helped to establish that such mutants arose spontaneously and not in response to an environmental cue.

Esther's Invisibility

In 1958, Joshua Lederberg received half of the Nobel Prize in Physiology and Medicine "*for his discoveries concerning genetic recombination and the organization of the genetic material of bacteria*" (the other half was shared by George W. Beadle and Edward L. Tatum). In his acceptance speech (5), Joshua recounted how he had enjoyed the company of many colleagues, especially that of his wife. He referred to the replica plate method and the F factor, but without mentioning the role of Esther in both discoveries. The only mention of her work was her 1950 discovery of lambda, which she had published as the sole author. Luigi Cavalli-Sforza said of Esther in 1974 that the long collaboration with her husband prevented her from having a stable and independent job, something that she fully deserved (11)].

Her ex-husband Joshua's website in the Profiles of Science Web of the U.S. National Library of Medicine had not a word about Esther and did not even mention her death.

Mercé Piqueras is a freelance science writer and science editor with a background in microbiology, based in Barcelona, Spain. She was President of the Catalan Association for Science Communication (2006-2011) and Vice-President of the Catalan Society for the Society for the History of Science and Technology (2009-2011). She is coauthor of a scientific guide to Barcelona (*Walks around the Scientific World of Barcelona*, Duran X and Piqueras M, Ajuntament de Barcelona, 2003).

References

1. **Lederberg E.** 1950. Lysogenicity in Escherichia coli strain K-12. *Microb Genet Bull* **1**:5–8.
2. **Lederberg E.** 1951. Lysogenicity in E. coli K-12. *Genetics* **36**:560.
3. **Lederberg J, Lederberg EM.** 1952. Replica plating and indirect selection of bacterial mutants. *J Bacteriol* **63**:399–406.
4. **Lederberg J, Cavalli LL, Lederberg EM.** 1952. Sex compatibility in Escherichia coli. *Genetics* **37**:720–730.
5. **Lederberg J.** 1959. A view of genetics. Available at the web of the Nobel Foundation http://www.esthermlederberg.com/Clark_MemorialVita/HISTORY52.html (visited: 25.10.2013).
6. **Muniesa M.** 2011. Lambda bacteriophage, a model of genetic decision (In Catalan) In: Corominas M, Valls M (eds) Organismes models en biologia. *Treballs SCB*, **62**:19-30.
7. **Raymann K, Bobay LM, Doak TG, Lynch M, Gribaldo S.** 2013. A genomic survey of Reb homologs suggests widespread occurrence of R-bodies in proteobacteria. *G3* (Genes Genomes Genetics) **3**:505-516.
8. **Richmond C.** 2006. Esther Lederberg. Obituary. The Guardian http://www.theguardian.com/science/2006/dec/13/obituaries.guardianobituaries (visited: 25.10.2013).

9. **Vandamme EJ, Miedzybrodzki R.** 2013. Phage therapy and phage control: ...to be revisisted urgently! *J Chem Technol Biotechnol* 10.1002/jctb.4245.(published online before printing).

10. http://www.esthermlederberg.com/Clark_MemorialVita/HISTORY52.html.

11. http://www.estherlederberg.com/LLCS%20Cavalli%20testimonials.html.

History of Science and Technology (2009-2011). This article first appeared in SEM@foro, a publication of the Spanish Society for Microbiology, and is being reproduced in translation by kind permission. The original article is available for download:

bit.ly/1PB9jSf

July 28, 2014

bit.ly/1GQjbQ1

#96

by Elio

Bacterial cells are smaller than eukaryotic cells, with some exception. Viruses are smaller than bacteria, with some exception. These are just two examples of why definitions in biology, unlike in physics and chemistry, most often need to be qualified. What does this tell us about biology?

February 28, 2013

bit.ly/1GQ88q9

16

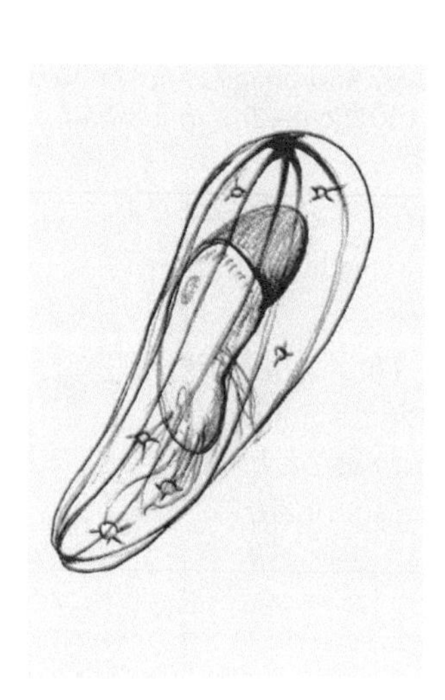

How Proteomics Got Started

by Frederick C. Neidhardt

The following is a personal account of the beginning years of what is now called *proteomics*. Odd that I, almost a Luddite, should be writing about the origin of a field initiated by a dramatic technical advance; I tend to avoid complex new scientific instruments and techniques. As a graduate student under Boris Magasanik at Harvard Medical School during the early 1950s, I was glad that my project (induced enzyme synthesis in bacteria) could readily be approached with simple technology. Bacterial growth could be monitored turbidimetrically with a Klett colorimeter; the same instrument could provide colorimetric assays of enzyme activities. Only the phage geneticists of that era, using sterile toothpicks to pick viral recombinants or mutants from plaques on Petri dishes, had it technologically easier.

Around me at that time in Harvard University's Department of Bacteriology and Immunology (now Microbiology and Molecular Genetics) were gifted individuals who on occasion were forced to purify proteins using laborious and personally onerous techniques. Not a life for me, I decided, even though H. Edwin Umbarger assured me that purifying an enzyme "*developed character*."

Beside laziness, there was a second, more fundamental, reason I never purified a protein. Cell growth was the biological event that had hooked me as a graduate student, and work that began by smashing cells into little bits seemed inappropriate.

Nevertheless, within the next six years I would find myself absorbed in two major aspects of cell growth physiology that involved proteins, and these subjects would prove more intractable than the purification of proteins. *Catabolite repression* (or, more generally, how bacterial cells choose to utilize multiple carbon sources) and *growth rate modulation* (how bacterial cell size and composition are interrelated with growth rate) were two processes directly related to cell growth rate.

The comprehensive work of Schaechter, Maaløe and Kjeldgaard on *Salmonella* cell growth rate and composition appeared shortly before my studies with *Klebsiella aerogenes*. Fundamental laws of bacterial growth were established by these studies in the early 1960s. Nevertheless, these laws were supported only by observation and by the easy rationale of their selective value to the cell; they were bereft of biochemical explanation. Both catabolite repression and growth rate modulation proved to be fascinating, but vexing; only now, fifty years later, are these processes approaching mechanistic solution.

A major reason for their intractability lay in limitations in our ability to approach the living bacterial cell. For most of the twentieth century the study of the physiology of bacteria (and, indeed, all other organisms) was largely reductionistic. The living cell was taken apart and studied biochemically, or was dissected by the increasingly powerful marriage of biochemistry and genetics. The triumphs of this approach were notable.

Still, catabolite repression and growth rate modulation joined a list of questions that could not be answered by the reductionistic approaches of biochemistry, even when augmented by the power of genetic analysis. Questions of the following sort had to be postponed (or were never asked) because the tools to approach them were not available:

- Why don't bacteria of a given species grow at the same rate on all carbon and energy sources?
- Is there a growth rate-limiting step during steady state growth of a bacterial culture?

- How many changes take place in a bacterium transitioning from growth to non-growth?
- What causes the size and macromolecular composition of a bacterial cell to be much more dependent on its rate of growth than on the chemical nature of its food?
- How do bacterial cells prioritize their choice of food when given options?

By the mid-1970s my mind, filled with unanswered questions about growth physiology, was searching for a new way to approach the bacterial cell. That way was revealed, not by anyone in my laboratory, but by a graduate student named Patrick O'Farrell at the University of Colorado at Boulder. A postdoctoral fellow in my laboratory at the University of Michigan, Steen Pedersen, one of the keenest of disciples of Ole Maaløe in Copenhagen (and one of his most honest critics) returned from a visit to Colorado in 1974 and reported to our laboratory that a graduate student there had produced a two-dimensional polyacrylamide gel system that could resolve the proteins of a bacterial cell on an array that looked as cool as "*the sky on a starry night*."

Steen's information electrified us, for we realized that a fundamentally new approach to bacterial growth physiology had become possible. Instead of asking the cell for information about a protein of interest to us, *we could finally interrogate the cell about the proteins that were important to IT* in any given situation. The cell could now reveal to us what lay behind the biological Green Door (in reference to an infamous American pornographic film of that era). For the first time the road to a global analysis of cell physiology could be imagined. And, in retrospect, it is clear that the era of proteomics began in 1975, the date of publication of Patrick O'Farrell's thesis research in the *Journal of Biological Chemistry*. His paper was quickly recognized by a variety of molecular biologists as a true technological breakthrough. Citations in the next 30 years numbered over 16,000 (in spite of the fact that the manuscript was initially rejected with two disparaging reviews which had to be overruled eventually by members of the journal's editorial board).

For the first time we could now learn what the cell had to teach us about its complement of proteins and about adjustments to different environmental conditions. This new ability to listen to the cell led soon to new insights into growth rate physiology. But before this could happen it was necessary to add several features to the O'Farrell technique.

First, we recognized that we had to standardize the two-dimensional gel system of O'Farrell in order to compare the protein arrays from different samples. This required extreme attention to details of procedures and quality of reagents. The genius of O'Farrell's system was that it employed two independent properties of proteins to separate them: their molecular weight and their isoelectric point. Isoelectric focusing in a gel tube containing ampholines to establish a pH gradient produced the first dimension—proteins lined up by their charge. Placing the resulting tubular gel on an electrophoretic gel slab containing sodium dodecyl sulfate, allowed the polypeptides previously resolved by charge now to be segregated by their size. The resulting two-dimensional polyacrylamide gel (2-D gel) was then stained and dried for subsequent inspection. A beautiful picture—but to be useful, 2-D gels had to be reproducible, and this was not an easy task for a number of reasons. In the end it took years of perfecting sample preparation and gel casting (not to mention improvements in ampholines) to get to the stage where computer-driven pattern matching could align a whole series of "starry patterns" from the multiple samples of an experiment.

Second, once the pattern-matching problem was in hand (no small feat), the issue became one of accurate measurement of the quantity of protein in the individual spots across the gel set. Clever uses, first of isotopes, then of differentially colored samples, were devised to obtain reasonable quantification. As a result, it became possible for the cell to display much of the array of changes made in its proteome (the totality of its several thousand proteins) as the cell adapted to its environment.

Fortunately, these tasks of standardizing and quantifying O'Farrell gels were approached by many individuals skilled in scientific technology. James Garrels at Cold Spring Harbor Laboratory, Norman G. and N. Leigh Anderson at Argonne National Laboratory, and Julio Celis at the University of Aarhus, Denmark, were some of the people who early on used their considerable skills to expand the usefulness of 2-D gel technology.

But still a third attribute had to be added to 2-D gels for maximum usefulness: the identities of the "starry" spots on the gels had to be determined. For the bacterium *Escherichia coli* and its close cousins, my laboratory in Ann Arbor mounted a full-scale effort to correlate spots on the 2-D gels with known proteins. Hundreds of protein spots were identified through the use of purified proteins (donated, naturally, by others) and mutants in known genes. Everyone in my laboratory contributed to this effort; unfair as this is, I'll single out only two because of their germinal work in identifying spots and because of their tireless energy in teaching the 2-D gel process to all the others: Ruth A. VanBogelen and Teresa Phillips.

Needless to say, the identification of spots might be regarded as tedious drudgery—and it was — save for the thrill that we were simultaneously making discovery after discovery using the 2-D gels: heat-shock and cold-shock proteins, proteins under stringent control, proteins that vary monotonically with growth temperature, proteins that vary with growth rate—and we were not simply learning which proteins exhibit a certain behavior, but what fraction of the cell's proteome was involved in different physiological responses to stress or starvation. These discoveries led Ruth VanBogelen and her colleagues to the concept of *protein signatures*. A protein signature is the set of proteins that, by their amplification or suppression, signal a particular physiological stress state of the cells. One learned how to recognize when a cell was in a state of energy starvation, or oxidative stress, or membrane damage, or... the list goes on. One can imagine the gigantic usefulness of this approach when a pharmaceutical company is exploring how a potential therapeutic agent acts.

But we should bring this story to a close quickly, because from the mid 1990s onward the explosion of cell protein technology transformed the field from what Pat O'Farrell had created to one with a formidable arsenal of techniques for protein resolution and measurement. The term *proteome* was introduced in 1996 to refer to the totality of proteins in a cell, and this quickly gave rise to the noun, *proteomics*, to designate studies of the proteome. The 2-D gel technique introduced by Pat O'Farrell has inspired others to develop improved techniques for monitoring the global pattern of a cell's total protein complement. The availability of DNA sequences with reasonably accurate annotations, for the genomes of hundreds of species has made it possible to develop separation techniques that enable tandem mass spectrometry to provide the "second dimension" to primary fractionation procedures, and as a result, enable protein identifications an order of magnitude beyond that which was achieved in the first two decades of the 2-D era.

To be sure, the current armamentarium of proteomics is being used in highly targeted ways to explore previously identified sets of "proteins of interest" (as our law enforcement agencies might call them), but I want to emphasize that Pat O'Farrell's development of the first method of spreading out the proteins of a cell was at the start, and particularly for me, the initiation of an exciting new grammar of scientific questioning.

Frederick C. Neidhardt is F.G. Novy Distinguished University Professor, Emeritus, Department of Microbiology and Immunology, University of Michigan Medical School at Ann Arbor.

December 21, 2009
bit.ly/1NlhOst

#31

by Elio

Many unicellular protists have a very complex body plan. One can find "eyes," "legs" (see our posting on Euplotidium), "mouths," food vacuoles, etc. Some of these structures reflect the feeding habits of the organisms. The question arises, why haven't they have become multicellular?

April 17, 2008

bit.ly/1ZSjF3X

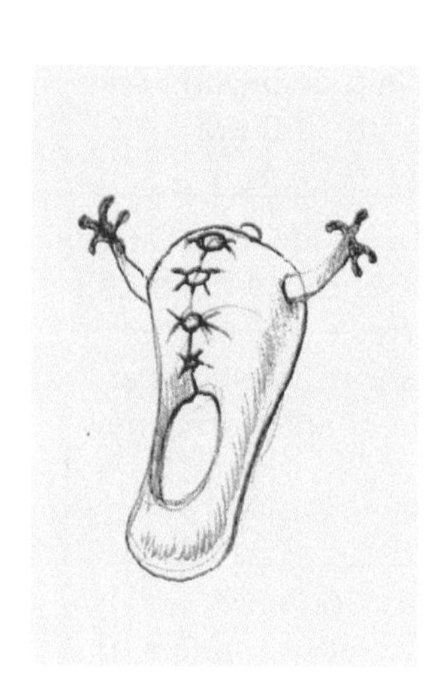

17

Did van Leeuwenhoek Observe Yeast Cells in 1680?

by Nanne Nanninga

It is common knowledge that beer was produced by the ancient Egyptians and that van Leeuwenhoek (1632–1723) was the first to see yeast cells. However, what was defined as yeast in the seventeenth century is different from that of today. So did van Leeuwenhoek really observe yeast? In attempting to answer this question it might be helpful to describe some fundamental work on yeast by Charles Cagnard-Latour (1777–1859) published in 1838. (Recall that the cell theory dates from 1839.) This tells us of the beer brewing and wine making state of the art around that time. It was known that the addition of yeast to properly treated grains of cereals would produce alcohol and carbon dioxide from the extracted malt sugars. In the times of van Leeuwenhoek, yeast was considered an inanimate paste with no connection to living cells. What did Cagnard-Latour see? It is important to use the term "see" because, as he emphasized, he wished to approach yeast research in a new way, that is, by employing a microscope. (This approach was also followed by F. T. Kützig and Th. A. Schwann at the same time.)

Cagnard-Latour went to an English-style brewery in Paris where every hour he took a sample for study from the container in which the beer was being produced. Initially, he observed globules of different sizes; later he saw small vesicles protruding from the globules that had arisen through budding (Fr. *par bourgeons*); and finally, after four hours, only doublets were present. He concluded the initial small vesicles had increased in size through growth. If so, the doublets should not be adventious aggregates. To test this he forcibly shook his microscopic preparation with a kind of thorn (Fr. *poinçon*) and observed that the doublets remained intact. Remarkably, he noticed that upon dissolution of the doublets a scar remained at a variable place on the globule's surface. He used the terms *cicatrice* (Fr. for scar) and *marque ombilicale* (Fr. for navel).

The number of globules increased in time, as did their total weight. His general conclusion was that beer brewing or fermentation involves the growth and proliferation of an organism. He also observed that commercial yeast contains singlets exclusively. So at this stage yeast was redefined, changing from an inanimate to an animate object. But was it an animal or a plant? Here he used the same approach as van Leeuwenhoek. The "little animals" of the latter were considered animals because they exhibited autonomous movement. Because yeast did not, Cagnard-Latour classified his growing globules as plants. Cagnard-Latour made another observation that I find highly relevant to the interpretation of van Leeuwenhoek's data. He saw his globules emit gas bubbles which caused individual globules to rise to the surface of the fermenting fluid in the fermentation vat.

Now, what about the observations of van Leeuwenhoek?

According to Cagnard-Latour in a footnote to his paper of 1838, van Leeuwenhoek considered yeast to be derived from flour. When presenting his work in Paris to the Academy of Sciences in 1837, he mentioned that he had not been aware earlier of van Leeuwenhoek's work. However, the first sentence in van Leeuwenhoek's letter of June 14, 1680, directed to Mr. Thomas Gale (1635/1636-1702), a fellow of the Royal Society, summarized his results in quite a straightforward manner: "*I have made several observations concerning yeast and seen throughout that the aforesaid consisted of globules floating through a clear substance, which I judged to be beer...*" Van Leeuwenhoek continues the above sentence by stating: "*...in addition I saw clearly that every globule of yeast* ("gist" in Dutch) *in turn existed of six distinct globules and that just of the same size and fabric as the globules of our blood.*" Note that the diameter of a red blood cell is roughly 7 µm, which is comparable to the size of a yeast cell.

Continuing his letter, van Leeuwenhoek reports that when sampling beer being brewed he observes in the fluid "*many small particles*" ("*seer veel kleyne deeltjens*") which cause the beer to appear turbid. "*Some particles were round, others irregularly shaped and some were bigger than others and they seemed aggregates of 2, 3 or 4 of the said particles and others again of 6 globules and the latter constituted a perfect globule of yeast.*" He interpreted this as aggregation (*stremming* in Dutch) and made many vain efforts to record the aggregation process. (Recall that Cagnard-Latour did the opposite: he attempted to demonstrate that yeast doublets arise from singlets.) To further appreciate the spatial organization of the yeast globule (the sextet), van Leeuwenhoek made a model composed of 6 wax globules which he arranged in such a way that all 6 globules could be seen when viewed from above. Clearly, this is an early, if not the earliest, example of model building for the interpretation of scientific data.

Van Leeuwenhoek speculated about the origin of the globules. (Note that he denoted single particles as well as sextets as globules.) "*Flour or meal of cereals is composed of globules which are first dehydrated by heat and when united with water which we are allowed to call beer, very small particles in the beer aggregate upon cooling to form a bigger particle, the singlet globule. The singlets aggregate to form a sextet, the yeast globule.*" (I did not translate this literally to keep it readable.) In other words, according to van Leeuwenhoek, yeast arises out of cereal flour or meals during brewing and as such yeast is not a distinct biological entity.

We are now approaching the answer to our main question. As Cagnard-Latour did later, Van Leeuwenhoek also observed air bubbles coming to the surface of the beer in the brewing container. He calls them "air" because, unlike Cagnard-Latour, he was unfamiliar with gaseous carbon dioxide. To his regret, he was unable to trace the origin of the "air." Then follows a very striking intermezzo which starts with: "*But what shall we say of the numerous air bubbles that were produced from lobster eye, when we lay them in vinegar?*" In summary, he noted that if enough air bubbles remain attached to pieces of lobster eye, the latter float to the surface of the vinegar. When the bubbles dissolve in the air, the pieces of lobster eye sink again. This process repeats as long as air bubbles are produced. Surprisingly, van Leeuwenhoek makes no explicit connection to yeast. The term *lobster eye* (*kreeften-oog* in Dutch) in the letter of van Leeuwenhoek is not clear-cut. It seems also to refer to a material substance (a *stoffe* in Dutch), which affects the salinity of an acidic solution.

This lobster story reminds one of the observations later made by Cagnard-Latour on yeast, as described above. But why did van Leeuwenhoek describe his earlier microscopic observations on fragments of lobster eye in his letter describing his observations of yeast globules? Presumably because, I believe, he was inspired by the behavior of yeast in a brewing container. Thus, irrespective of the identity of lobster eye, a reasonable answer to our main question is that van Leeuwenhoek did indeed observe growing yeast cells, despite the fact that he could not place them within a biological framework.

Author's Note: *Translations of van Leeuwenhoek's Dutch to English are mine, and thus I am responsible for any errors.*

Nanne Nanninga is Emeritus Professor of Molecular Cytology, Swammerdam Institute for Life Sciences, University of Amsterdam, The Netherlands.

Primary sources for this contribution include the 1680 letter of van Leeuwenhoek that has been kindly brought to my attention by Warnar Moll. The reference to the 1838 paper of Cagnard-Latour I found in the elegant 1871 paper of Thomas Huxley on yeast. Also highly recommended is a scholarly review on 19th century "biochemistry" by Herbert C. Friedmann. Very informative is the review of James A. Barnett.

References

Antoni van Leeuwenhoek. Letter to Thomas Gale of June 14th 1680, pp. 6-10 in the second printing carried out by Hendrik van Croonevelt at Delft in 1694.

Cagnard-Latour C. Mémoire sur la fermentation vineuse: présenté á L›Académie des Sciences, le juin 1837. Comptes rendus des séances de l› Académie des Sciences, l›octobre 1838.

Cornish-Bowden A (ed). 1997 p 67–122. *In Yeast. The Contemporary Review (1871). Collected Essays VIII.* ***Herbert C. Friedmann.*** *From Friedrich Wöhler's urine to Eduard Buchner's alcohol. From Beer in an Old Bottle: Eduard Buchner and the Growth of Biochemical Knowledge.* Universitat de Valencia, Valencia, Spain.

Barnett JA. 2003. Beginnings of microbiology and biochemistry: the contribution of yeast research. *Microbiology* **149:**557–567.

April 8, 2010

bit.ly/1klHD7k

18

The Louse and the Vaccine

by Elio

Several events in my life that I will recount below have combined to make me feel connected to the rickettsiae. Consequently, I was drawn to a recent book somewhat bombastically entitled *The Fantastic Laboratory of Dr. Weigl* by Arthur Allen. It centers on an astonishing microbiological story that took place before and during World War II, centered around the making of an effective typhus vaccine. This vaccine became an instrument of war, and its production involved intrigue, deception, and, in great measure, heroism and high class science.

One of the great concerns of military physicians in time of war has been the spread of epidemic typhus among the troops, a high-mortality disease that can have disastrous consequences. The rickettsiae that cause this disease are bacteria, intracellular in humans and lice, transmitted by the body louse (but curiously, not by the everyday head louse). Body louse infestation is the hallmark of poverty, lack of hygiene, and crowding. In other words, it's a companion of war combatants. As the Germans and Russians in 1939 invaded Poland, where typhus eventually became prevalent, they became fanatically worried about the spread of the disease and went to great lengths to obtain vaccines that would prevent it. Successful anti-typhus vaccines were not readily available due to difficulties in working with the rickettsia. In particular, these intracellular bacteria do not grow on artificial lab media, and cell culture techniques had not been developed sufficiently to be of use for growing them. Not only that, most laboratory animals are relatively unaffected by the human agent of typhus and are thus not capable of providing large amounts of vaccine. Besides vaccination, the other available control measure was delousing. Although this was considered an effective way to control typhus and was done on a large scale, it could not always be carried out under wartime conditions. In time, the American Army found that fumigating people with DDT was indeed an efficient way to get rid of lice and control typhus, but that came much later.

The most successful anti-typhus vaccine of the 1930s had been developed by Rudolf Weigl, a Polish investigator and biology professor whose name is in the title of Allen's book. He was a well-established scientist at the University of Lwów (now Lviv, by default), and had been repeatedly nominated for the Nobel. At the time, Lwów was a major center of learning, science, and culture. The city was successively occupied by the Russians (who slaughtered 14,000 Polish officers in Katyn Forest and deported untold more to Siberia) and then the Germans. In 1944, it was again taken over by the Russians, who did some thorough ethnic cleansing and thus converted what had basically been a Polish city (it had been 62% Polish and 24% Jewish) into a then-Russian/Ukrainian one. Through all this tumult, Weigl managed to run an efficient and productive lab, centered during the war years on the production of his vaccine. He was allowed to carry on and run his lab because of the high priority the Germans and Russians put on the vaccine effort.

The Weigl vaccine was a genuine tour de force. Lacking regular experimental animals in which to grow the rickettsiae, Weigl turned to lice, in which these microorganisms grow abundantly. But for that you need a louse colony. How to establish one? Get people to volunteer to strap on their legs little lice-containing cages and let the lice bite them and suck their blood for an hour, which provides the required blood meal. The "feeders" included a number of intellectuals and scientists, some Jewish, who survived because the Germans deemed them essential to vaccine production.

To make the vaccine, healthy lice were then given an enema of a rickettsial suspension, administered by a steady-handed "inoculator" using a very thin cannula. Skilled "inoculators" could inject up to 2000 lice per hour. After a week or so, the germ-laden louse midguts were harvested and ground in Weigl's mortar. Phenol was added to the suspensions, and, a few manipulations later, the vaccine was ready. Details of the preparation can be found in an article (bit.ly/20943sY) by Krynski et al.

Recently, I had the good fortune to have a most informative phone conversation with Waclaw Szybalski, an eminent synthetic molecular biologist and professor at the University of Wisconsin whom I have known for many years. Szybalski, who has written a fine article on Weigl, figures prominently in the book. Among other reasons, as a young man of 19 to 23, he had been in charge of the feeding operation of Weigl's louse colony. He describes how this was done in this article. His brother Stanislaw, 6 years younger, had worked there as a "harvester" or "preparator" collecting the midguts of the infected lice. Both Szybalskis told me that they attribute their survival to their job in the Weigl lab.

A particularly touching aspect of the story is that the Weigl lab produced vaccine lots of different potency, the stronger ones being given to resistance fighters and smuggled into ghettos, and the weaker ones supplied to the German Army. All this was done under the probing eyes of the Germans; thus, it involved huge risk. Had this ruse been discovered, it would have meant certain death. According to Szybalski, among those employed in the Weigl Institute and thus sheltered were members of the WWII Polish Underground (Armia Krajowa or AK), including members of a branch known as "Zegota," the only secret organization in Hitler-occupied Europe especially dedicated to protect Jews. For his contribution to saving Jews, Weigl was named "*Righteous among the Nations*" by Israel.

Along with the vaccine story, another one involving both microbiological savvy and extreme courage unfolded in a small town in southwestern Poland. There, two young physicians, Edward Lazowski and Stanislaw Matulewicz, devised a way to shield both the local persons from being sent to Germany as slave laborers and Jews from certain death. The two took advantage of a serological quirk, namely that rickettsiae and strains of a fairly ordinary bacterium, *Proteus vulgaris*, happen to share a common surface antigen. For this reason, detection of antibodies to *Proteus* could be used as a diagnostic test for typhus. This *Weil-Felix test*, so-called for its discoverers, has fallen into disuse because it's neither sensitive nor particularly specific. However, at that time it was the handiest way to diagnose typhus cases and was widely used. The two physicians figured out that injecting killed *Proteus* cells into people would render them Weil-Félix positive, even though they didn't have typhus. When they dutifully sent the sera of these people to German labs, the reports came back as typhus positive. Thinking that the village was a hotbed of typhus, the Germans declared it an epidemic zone and therefore off-limits, which saved its inhabitants from harm. Consider the courage involved: had the Germans examined any of these patients, they would have found them to be free of typhus, and thus would have uncovered the ploy, with horrible consequences.

This remarkable story, which inexplicably is not mentioned in the book on Weigl, took place in a village about 15 miles from Tarnobrzek, which is the town in southeastern Poland where my mother's family had lived for many generations until leaving during the First World War. I traveled to that town a couple of years ago but was unaware of the heroic event that had taken place in the vicinity. So this is one of my personal "rickettsia threads."

But there is more: just after getting my PhD at the time of the Korean War, I was assigned to the rickettsiology lab at Walter Reed in Washington, DC. There, as a private, I was given the order to settle once and for all the question as to whether rickettsiae were more like bacteria or more like viruses (or quaintly, "something in-between"). This puzzle had persisted because rickettsiae did not grow on media that supported the growth of bacteria but only within host cells. I obediently followed the fate of individual rickettsial cells in cultured cells under a microscope and, lo, I beheld that several had divided in two. Not only that, we later determined that they possess bacteria-like cell walls, both of which findings settled the question of rickettsial identity forever, placing them snugly in the bacterial fold.

You would think that these various encounters would have turned me into a rickettsiologist, but that was not to be. The lure of *E. coli* and *Salmonella* was just too great for me in my early days. Now, I return to these stories as to a familiar novel from youth—with fascination at the levels of detail and new meaning that have emerged with each reading.

August 24, 2015

bit.ly/1NRTFBY

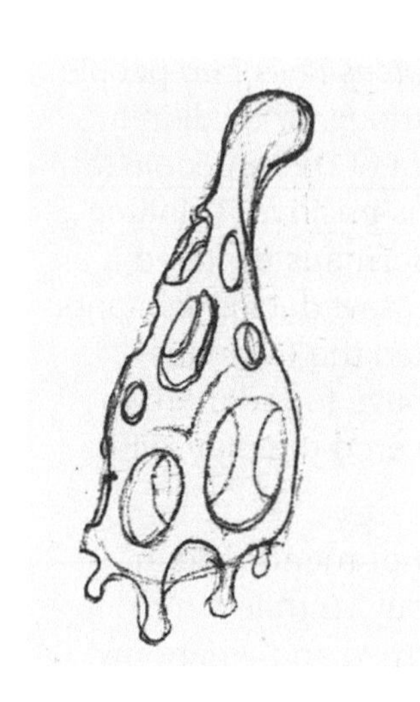

19

The Ten-Minute Leeuwenhoek Microscope

by Patrick Keeling

I was on leave from teaching for a couple of years. The summer before re-starting my third-year Protistology course I began to think about some things I wanted to change. One thing I wanted to do was to add a section on the history of microbiology to put things into perspective and hopefully connect students with the material a bit. My colleague, Max Taylor, had a replica of a van Leeuwenhoek microscope that he once showed to me, and I thought it would be fun to make one as close to the original design as possible to show the class what it was like. I did some superficial snooping around about how it was built, including finding the original paper in the *Proceedings of the Royal Society* where he described the design. I realized it would be pretty straightforward, including making a lens that was pretty close to the ones he would have used.

Deep down I knew that this was mostly to find some excuse to buy an anvil because, seriously, what kind of guy does NOT want to have his own anvil? So, one anvil, a big hammer, and a bit of brass later, I had made all the components for two Leeuwenhoek microscopes. I later found a superbly detailed web page describing how to do it that would have saved a lot of trial and error (especially my mistake in threading both holes in the L-shaped piece that connects the microscope plate to the stage), a source that I would recommend to anyone who wants to make one of these.

One of the big surprises for me was how easy it was to make the lens without any grinding. Again, there are loads of web pages with great instructions to make one of these bi-convex lenses (e.g., the one linked above), and I recommend them for anyone serious about wanting to replicate one of these microscopes. But it also got me to thinking: my idea of showing a replica microscope to my class would be all right, but it would be way better still to have them actually *make* and *use* one themselves. The lens is the intimidating part, but that was easy, so what about the rest of the microscope?

It is immediately obvious that the plates that sandwich the lens can be made of stiff paper, but devising a mechanism to hold the sample and focus it was harder. I tried a few solutions that were overly complex, not to mention failures, but then had an idea to use sticky tack. This is the gummy stuff used to hold posters up on the wall. It is not only sticky, but also has great elastic properties, so you can use it to mount all kinds of samples in front of the lens and also to pivot the sample away from

Patrick melts a glass tube (while looking away!) to make a Leeuwenhoek microscope at the UBC Advanced Molecular Biology Labs High School Science Teacher Conference (October 2008).

Credit: Patrick Keeling.

or towards the lens. This, in a stroke, eliminates the two biggest design problems with a paper microscope, and I think that if Leeuwenhoek had sticky tack in the seventeenth century, he would have used it, too.

In any case, I made a few of these and they all worked well and were easy to put together, so I set up the first three labs in my class to tie into lectures about the history of microbiology and taxonomy. The students made their own microscopes, measured the size of their lenses, and calculated the power, thus learning something about optics. They then used them to observe natural samples. After that, they observed the same samples with the 1960s technology at their disposal in the lab, so they experienced how it felt for the pioneers of microbiology.

All of the students made decent microscopes. Indeed, this class has a lab exam at the end of the first term, and for one of the "stations" during the exam they were not given a microscope. Instead, they had a fixed slide, a Pasteur pipette, some paper, and a Bunsen burner. They were given 10 minutes to make a microscope that was good enough to identify the protists on the slide. Happily, the students aced this question, so technically this project is pretty accessible.

It also seemed pretty popular with the students. A little while later I was asked to demonstrate the lab to some public school teachers at a professional development day organized by David Ng and Joanne Fox at UBC (who posted a couple movies from the day on YouTube). Myself and the TA from my class, Gillian Gile, taught 90 teachers how to make one, and again it seemed pretty popular. Some teachers have already used the exercise in their classes, and a couple reported back that it worked well with students of a broad range of ages. One teacher even told me that she was going to include protistology in her curriculum in the future so the lab had better context, which I thought was great. I have also had a couple students demonstrate this directly to public school students and report back that it was popular and successful.

Overall, this has been a great experience. I find that getting people interested in microbiology and microscopes is challenging, I think because the microbial world is so abstract to them. By delving into the history, the students experienced a fascination akin to that felt by many young students in physics when encountering the history of its pioneers. But being able to build a sixteenth century microscope yourself is better still (and although you don't need an

anvil, you still get to play with fire and use a drill, so it is a lot of fun!). I would recommend trying this with any class or lab from high school up to senior undergrads, and if you do please let me know how it goes, or if you can suggest any improvements.
I made a set of instructions and posted them on my web page (bit.ly/1RAd5ZN), so anyone should be able to quickly make one of these, and teach their class how to do it too.

Patrick Keeling is a Senior Fellow of the Evolutionary Biology Program of the Canadian Institute for Advanced Research (CIFAR), Director of CIFAR's Integrated Microbial Diversity Program, and Professor, Department of Botany, University of British Columbia.

July 20, 2009

bit.ly/1RTnm3V

#52

by Mark Martin

Given the ubiquity of intracellular associations between eukaryotes and prokaryotes, why are there so few reports of prokaryotes living "within" prokaryotes?

August 20, 2009

bit.ly/1GQ7Xem

20

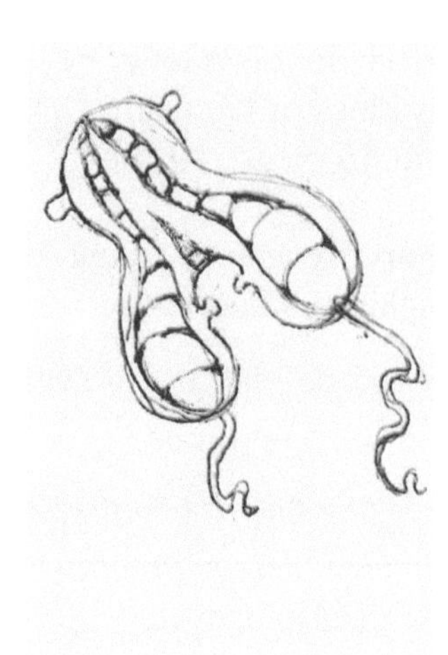

Microbiology in the Andes: Ancient and Unexpected

by Elio

Thanks to the investigations by the Ecuadorian physician and scientist Dr. Byron Núñez Freile, I learned of a surprisingly high level of scientific development that took place long ago in a remote region of the world. Quito, the present-day capital of Ecuador, is nestled amidst the high Andes and was the northern capital of the Inca empire. It was conquered by the Spaniards in 1534. In this exceedingly distant land, Jesuits established a college within a year of their coming in the late sixteenth century. By 1622, they founded one of the oldest universities in the Americas, the Universidad de San Gregorio Magno. This was earlier than the founding of Harvard, which happened in 1642. With the passing years, the two universities may not have enjoyed a parallel development, but early on they were likely of comparable quality. Soon, San Gregorio became a major institution with a most impressive library of 16,000 volumes, the largest in South America at the time. In its first thirty years of existence, the university granted 160 master's degrees and 120 doctorates, mostly in philosophy and theology. Nevertheless, the library holdings also included numerous scientific and medical treatises.

Quito can be reached by an easy flight these days, but in olden days getting to its lofty location (an altitude of 9,000 feet) required a week-long mule trek from the Pacific coast. Remote indeed! A major event of scientific relevance took place in 1736, when a geodetic French mission led by Charles-Marie La Condamine arrived, intent on measuring the circumference of the Earth at the equator. The French delegation interacted closely with members of the university, which resulted in a strong scientific legacy.

The scientific concerns of the times included the world we now call microbiology. No wonder. In 1589 a smallpox epidemic killed 37.5% of Quito's inhabitants. A description of the disease in a letter by one of the priests makes clear allusion to its contagiousness. Later on, several of the Jesuits made insightful observations about the etiology of infectious diseases. Among them was Juan Magnin (1701–1753), a Swiss missionary who became a member of the French Academy of Sciences, who stated: "*There are microbes that can only be seen with a microscope that are 27 million times smaller than the smallest that can be seen with the naked eye. These facts and others seem incredible.*"

And "*(the microscope) allows to establish that the dirt on the teeth is due to the accumulation of innumerable microbes; furthermore, it is likely that many of the diseases of the human body, especially leprosy and venereal diseases, are due to the accumulation of microbes.*"

Façade of the church La Compañía de Jesús, or the Company of Jesus (better known as Jesuits) in Quito, Ecuador, completed in 1765.

Credit: Tara McGovern.

The other major microbiological issue of those times, the theory of spontaneous generation, became the concern of a native-born member of the faculty, Juan Bautista Aguirre (1725–1786). He wrote: "*I affirm...that the forms of animals, even insects, are not engendered by putrefactions but they arise from eggs or germs.*" He also stated: "*...with the aid of the microscope one discovers innumerable germs incredibly small in size, in the air, water, vinegar, blood, milk, etc. The most ingenious Leuvoiseck (sic) bore witness to having seen such small germs in a drop of water that 90,000 of them did not reach the size of a grain of wheat.*" What he lacked in spelling skills, he made up for by a good understanding of the literature!

There is more to this history. Soon after the Jesuits were expelled in 1767, the major Ecuadorian intellectual of that age, Eugenio Espejo (1747–1795), made important contributions to hygiene and the containment of smallpox. His words: "*Within the infinite variety of these living particles* ('atomillos') *we have an admirable resource to explain the prodigious multitude of diseases and symptoms...*" Born of an Indian father and a mestizo mother, Espejo was a notable polymath, a true product of the Enlightenment. Not only was he the most notable physician of his time in Quito, he was also a lawyer, a philosopher, and the founder of Quito's first newspaper. As a public figure, he laid the groundwork for the independence movement that eventually led to the liberation from Spain.

In the seventeenth and eighteenth centuries, Quito was a notable center of learning and discovery. Here, in splendid isolation, far from other universities and libraries, arose a sophisticated understanding of the world of microbes, both regarding their medical importance and their biological essence. This is nothing short of remarkable.

I confess to personal glee in this story. I spent my teen years in Quito and eventually became a student at the Universidad Central, the public institution that was built on the one founded by the Jesuits so long ago.

I am grateful to Dr. Núñez Freile for having brought this remarkable story to my attention. Dr. Núñez Freile is in charge of two highly informative blogs (in Spanish), one on infectious diseases and the other on hand washing.

July 19, 2010

bit.ly/1klHIrF

21

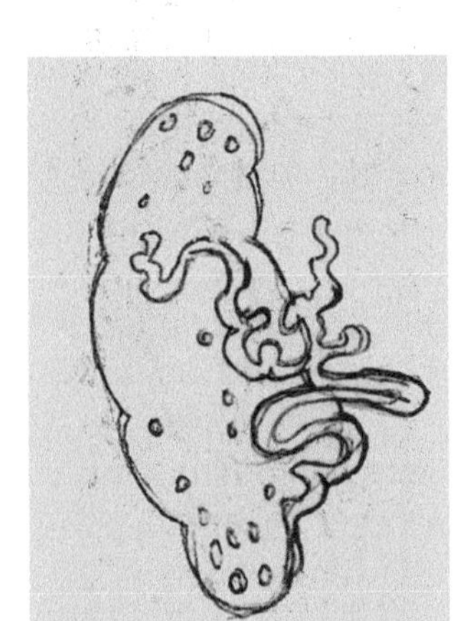

Dr. Rous's Prize-Winning Chicken

by Welkin Johnson

If one were to apply contemporary principles of evaluation to the work of Peyton Rous in his time and without the benefit of hindsight, this work would certainly be assigned a low priority, not fundable, and probably also not publishable in the trendsetting journals that today alone can confer recognition and prestige. Even in 1966, the year of the [Nobel] prize, it was impossible to guess at the full significance of the Rous sarcoma virus for human medicine and biology... The urgent lesson from the Rous experience, then, should be that it is the quality of the science that counts, not its compliance with a fashionable trend and not its perceived future value, which cannot be predicted.

–Peter K. Vogt (1996. *The FASEB Journal* 10:1559-1562.)

One hundred years ago today, Peyton Rous published the first in a series of papers describing experiments that began with work on a sarcoma from a single Plymouth Rock hen. In that initial paper, Rous showed that bits of the tumor could establish new tumors when injected into healthy chickens. A year later, in 1911, Rous published what has become a classic in the annals of virology—*A Sarcoma of the Fowl Transmissible by an Agent Separable from the Tumor Cells*—reporting that the tumors were transmissible from one chicken to another by injection of a cell-free, filtered homogenate of the tumor tissue. The conclusion, since confirmed a thousand times over, was that a virus caused the tumors. The virus, now known as Rous sarcoma virus (RSV), went on to play a starring role in twentieth century biomedical research.

Historical accounts tell us that Rous's findings were initially dismissed by many as a peculiarity of chicken sarcoma, with little or no relevance to human cancer. The view one hundred years later is far more spectacular. In the 1930's, reports of oncogenic viral agents began to emerge from laboratories studying cancer in inbred mice. Some of these cancer-causing viruses, like Rous sarcoma virus, had RNA genomes, and as a group they came to be known as the RNA Tumor Viruses. In 1958, Howard Temin and Harry Rubin published a description of a quantitative assay for studying Rous sarcoma virus in tissue culture. The assay took advantage of the rapid and reproducible way in which RSV transformed cells. Specifically, diluted solutions of virus could be applied to monolayers of cells, which were then overlaid with agar. Localized "foci" of transformed cells would then form. Virologists could count foci in a monolayer of cells the way bacteriologists count plaques on a lawn of bacteria, back-calculating to determine multiplicity of infection. Thus, by the 1960s, the study of RNA tumor viruses had become a mature field encompassing both *in vivo* animal studies and tissue culture-based, bench-top experiments. In 1966, Peyton Rous's contributions were belatedly recognized with a Nobel Prize.

Howard Temin's studies of RSV transformation in tissue culture, and in particular the observation that the transformed phenotype remained stable during passage of transformed cells, led him to the startling hypothesis that the RNA tumor viruses replicated through a DNA intermediate, or *provirus*. Although Temin published several papers in support of his provirus hypothesis, most of his contemporaries did not see the light until 1970, when he and David Baltimore independently proved that virions of RNA tumor viruses contained RNA-dependent DNA polymerase activity. This enzyme came to be known as reverse-transcriptase (RT). An opinion piece summing up the ensuing flood of follow-up studies was appropriately titled *Apres Temin, le Déluge*. In the wake of this flood, Francis Crick even felt obliged to defend the so-called central dogma of molecular biology. In addition to revolutionizing the study of RNA tumor viruses (subsequently called retroviruses), the discovery of RT was a watershed event in molecular biology, providing the means for generating cDNA and the key to reverse transcription–PCR (RT-PCR). In recognition of their work, Temin and Baltimore received the Nobel Prize in 1975.

By the 1970s several groups had begun zeroing in on the molecular nature of acute transformation of cells by RSV. Peter Duesberg and Peter Vogt reported that the RSV genome contained an extra bit of sequence not found in the genomes of very closely related, non-transforming viruses. In 1975, Dominique Stehelin, J. Michael Bishop, Harold Varmus, and Peter Vogt reported that this extra sequence was homologous to a host gene. The host gene, called *src* (from *sarcoma*) was the first proto-oncogene, and v-*src* (the version found in RSV), its oncogenic counterpart. Like RT before it, the discovery of viral oncogenes incited a dramatic shift in experimental emphasis, this time by giving scientists a molecular beachhead in the war on cancer. For discovering the cellular origins of viral oncogenes, Bishop and Varmus received the Nobel Prize in 1989.

It is noteworthy that during the sixty-six years separating Rous's chicken experiments and the discovery of *src* (1910–1976), no retroviral pathogens of humans were known (it wasn't until 1980 that Robert Gallo's group reported the isolation of HTLV-I, the first human retrovirus). Thus, seven decades of work on retroviruses had already had a profound impact on biomedical research when, in 1983, the RT-assay described by Temin and Baltimore helped Françoise Barré-Sinoussi and Luc Montagnier detect a new retrovirus in the cells of an AIDS patient. In 2008, Barré-Sinoussi and Montagnier were awarded the Nobel Prize for discovering the human immunodeficiency virus (HIV-1).

All thanks to Dr. Rous and his prize-winning hen.

Welkin Johnson is Professor and Chair of Biology at Boston College.

September 1, 2010

bit.ly/1OFxZJe

22

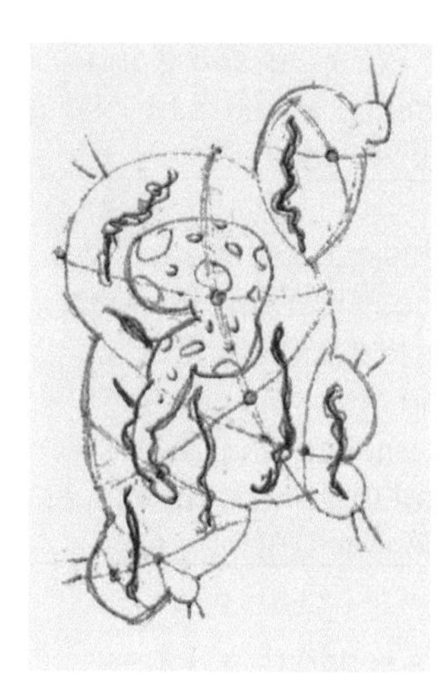

The Two Quantitative Steps in the Biology Growth Curve

by Elio

We are witnessing a highly influential influx of physicists, mathematicians, and engineers into biology. This is not the first time. Over the centuries, biology has been blessed by the involvement of people from other fields. During the last half century or so, we can point to two distinct incoming waves, each one bringing in many with different viewpoints and often greater quantitative skills. The first wave took place after World War II and played a key role in the development of modern biology. The second one is happening now.

In brief, I would like to compare these two episodes not as a historian, but, perhaps naively, simply as one who has lived through these periods. I started my graduate work in 1950, about the heyday of the development of molecular biology, and I am still around to observe what is happening now.

In the first wave, one could distinguish two streams of physicists heading into the field. One group retained its traditional roots in physics and used big machines to study biological phenomena. Modern radiation biology came of it, and although it provided magnificent tools such as isotopes, it was perhaps not as successful in answering basic biological questions as its proponents had envisaged. The other group started afresh, but with the typical physicist way of thinking. For the object of their study they looked for the smallest units. They zeroed in on bacteriophages, shunning the bacteria that they thought too complicated. Their tools were simple: Petri dishes and test tubes. One paradigm is the development of the phage *one-step growth curve* by Delbruck and Ellis, which permitted quantitative studies of viral development. Another was Luria and Delbruck's *fluctuation test*, which provided credible evidence for the random nature of mutations. The only math needed was the Poisson distribution.

Looking behind the historical facts for the moment, there is an obvious lesson to be learned here. As much as technology can contribute to a science, even greater is the value of novel approaches arising from people entering from another field. I can attest that working in one area impels one to continue to work along lines that one is familiar with. One wants to answer the "next question." It is unusual, although not unheard of, to take a serious pause and ask: "*What is the most important question in my field*?" Not so for newcomers. They have the freedom of starting out with such questions.

The current second wave has yielded an avalanche of studies often started by people with a physical, mathematical, or engineering background. I suppose one could divide these newcomers into two categories as well, but the distinctions are different from those in the first wave. One category is represented by people who deal with technology, but in innovative ways. They may have been encouraged by the great physicist Richard Feynman who, when asked how one should study biology, answered: "*Look at the thing*." Resulting from these efforts has been the stunning developments in optical, electron, and other microscopes. Sacred cows, such as the belief that the resolving power of the microscope is limited by the wavelength of light and the refractive index of the medium, have been left to graze in distant pastures. Likewise, cryotomographic techniques have allowed the electron microscope to be used on nearly undisturbed cells. I have commented on this Age of Imaging before.

The approach of the other category has been to make predictions of the quantitative sort, that is, models. This is their way of expressing the intricacies of a biological phenomenon in quantitative terms. A model is made and then it is tested against reality. It works in reciprocal ways. Sometimes the data used are already available, other times the models suggest experiments designed to test their plausibility, and sometimes both things are in play. For the student wanting to know more about this I cannot think of a better source that the textbook *Physical Biology of the Cell*.

We are now witnessing stunning developments in all aspects of biology. Just look at what you see on the lab bench. Test tubes give way to microfluidic apparatuses, optical tweezers leave ordinary micromanipulation in the dust, and so forth. And all this has practical relevance. Clinical diagnoses are beginning to be based on extremely rapid techniques such as magnetic resonance analysis, etc., etc. And new algorithms are piled upon algorithms with the gusto that characterizes bioinformatics. In brief, we are transforming our science from molecular biology to systems biology (complete with all the ambiguities of this new term).

The transformation of the landscape has indeed been dramatic. I cannot begin to guess where this will lead us, other than to be confident that unforeseen and unforeseeable discoveries will be made by the three-way collaboration of biologists, physicists, and mathematicians. Certain goals, such as being able to observe at atomic resolution individual molecules in their natural cellular environment, may take a while to achieve. But who's to say; maybe this, too, is just around the corner. And yet, have new laws of biology emerged? Is this a paradigm shift? In the Kuhnian sense, maybe and maybe not. Are we concerned with familiar questions, albeit with fabulous new tools, or are we entering into a different world of biological understandings?

March 22, 2012

bit.ly/1MAB9bc

23

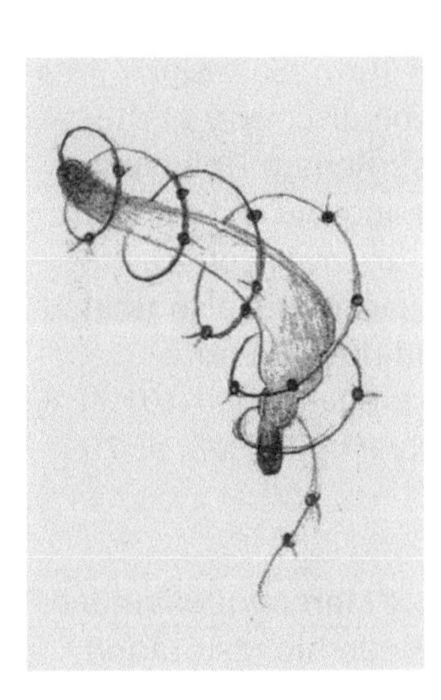

Tales of Centenarians

by Claudio Scazzocchio

Editor's Note: While it may be true that bacteria do not have neurons (at least not more than one per cell!), their modes of communication hint at the equivalent of a simple nervous system. But we need not apologize for the inclusion in this blog of a non-microbiological topic. The story here described has much poignancy and is relevant to all biologists.

The year 2012 was quite turbulent in Italian politics, but it ended with a display of near unanimity. The reason was the posthumous homage to the oldest Senator of the Republic and the oldest Nobel Prize winner, Rita Levi-Montalcini. In Italy, the president has the privilege of personally naming up to five senators–for–life, with all the legislative privileges and duties that entails. In 2001, President Ciampi appointed her to this post, which she held until her death at the age of 103 on December 30, 2012.

A School of Nobelists

In the wonderful course of Embryology and Histology given by Prof. Washington Buño in the School of Medicine in Montevideo in 1956, I acquired my taste for developmental biology through the description of classical experiments such as those of Driesch and Spemann. I also encountered the name of Giuseppe Levi, who had visited the country a few years before and helped set up the first tissue culture laboratory in Uruguay.

Among many excellent scientists, three future Nobel Prize winners came from the laboratory of Giuseppe Levi in Turin. Salvador (originally Salvatore) Luria got it in 1969 for the discovery of the mechanism of replication of bacterial viruses[1]; Renato Dulbecco in 1975 for discovering interaction of viruses and the cell's genetic material; and Rita-Montalcini in 1985 for the discovery of growth factors. A cousin of Rita-Montalcini, Eugenia Sacerdote, also trained with Giuseppe Levi, and eventually introduced tissue culture techniques and the polio vaccine to Argentina. She died at 101 years of age, Dulbecco at 98, and Luria, nearly a youngster, at 79.[2]

From Marengo to Munich

In 1800, Napoleon, the winner of the battle of Marengo, extended civil rights to Piedmont but, following the monarchic restoration, the Jews were not fully emancipated until 1848. After the creation of the Kingdom of Italy in 1861, the Piedmontese Jews became the vanguard of Italian Jewry. Most were Italian patriots and liberal. Some Italian Jews were nationalistic and eventually became fascist, even participating in Mussolini's 1922 "March on Rome." Others were active antifascists, like the brothers Rosselli, founders of the "Giustizia e Libertà" movement, who were assassinated in France by the regime's hit men. Giuseppe Levi, Levi-Montalcini's teacher, originally from Trieste, started as a Nationalist, volunteered as a doctor in the First World War, and eventually became explicitly and actively antifascist.

In 1938, Czechoslovakia was dismembered in Munich, while Hitler and Mussolini cemented their alliance, Mussolini now discovering his own anti-Semitism. In the 1938 "Manifesto of the Racist Scientists," it was proclaimed that the Jews were not members of the "Italian race." The Royal Decree #1779 followed and excluded Jews from all educational institutions. Giuseppe Levi had to give up his chair and laboratory in the Institute of Anatomy at the University of Turin. Luria followed the exodus of many others, going first to Paris until the debacle of 1940, then to Marseilles, Lisbon, and finally New York City. Dulbecco was drafted as a doctor and witnessed the massacre of his regiment on the Russian front and eventually became part of the anti-Nazi movement in Turin.

An Unorthodox Path

Rita Levi-Montalcini was born in 1908 to a well-to-do Jewish family. Her twin sister, Paola, with whom she remained close throughout her life, became an original and important painter, admired by de Chirico, while elder brother Gino was a distinguished architect. Overcoming family objections, Rita became a physician in 1936 along with Eugenia Sacerdote. Both joined Giuseppe Levi's laboratory as students. Rita and Giuseppe Levi continued their research in Belgium after 1938, soon to be interrupted by the German invasion. On the train back to Italy, Rita read a paper by Victor Hamburger (a refugee from Nazism to the USA who also died a centenarian at 101 in 2001). Hamburger showed that ablating the wing primordium of a chicken embryo resulted in the decrease in number of the corresponding neurons in the spinal cord. Both student and teacher repeated and perfected Hamburger's results in a clandestine improvised lab. In 1942, they published their results in an obscure Belgian journal. Hamburger proposed that the peripheral tissue stimulated the formation of the spinal cord neurons. The Italians, on the other hand, suggested that the peripheral tissue inhibits the death of the neurons.

In 1943, Italy collapsed, with the creation of a puppet fascist state in Northern Italy and the deportation of Jews. Eventually, Rita emerged from a clandestine refuge in Florence, where, after

Rita Levi-Montalcini.

Credit: Herme Castro.

the liberation of the town, she worked as doctor (and nurse) with the Anglo-American troops. Both she and Giuseppe Levi regained Turin and the laboratory in May, 1945. In 1947, Hamburger invited her for a short stay in his laboratory in St. Louis, Missouri, for the purpose of resolving their different interpretations of the ablation experiments. The initial six months became thirty years.

The Road to the Nobel Prize

The differences between their results were soon resolved in favor of the peripheral tissues inhibiting death of the pre-existing neurons. It is arguably the first description of programmed cell death. A former student of Hamburger, Elmer Bueker, had implanted sarcomas in chick embryos and saw that the "type 180" sarcoma became invaded by nervous fibers, with an increase in size of the nearest sensory ganglia. Rita and Hamburger, with Bueker's permission, repeated these experiments under a variety of conditions. In one experiment Rita implanted tumor fragments outside the embryo, in the chorioallantoic membrane. The results left no doubts on the diffusible, soluble nature of the nerve growth-promoting factor released by the tumors. Hertha Meyer, another refugee and ex-student of Giuseppe Levi, had set up a tissue culture lab in Rio de Janeiro. Rita flew to Rio carrying in her hand-bag two mice with implanted tumors. Co-culturing tumors and nerve ganglia led to an exquisitely sensitive assay of the not-yet-named nerve growth factor (NFG). Stanley Cohen joined the team, bringing to it biochemical know-how leading to the purification of NGF. An unexpected finding was that the salivary glands of male mice were a great source of NGF. In addition, Cohen found another novel factor, that for epithelial cell growth. But what was the physiological role of NFG? This was determined by injecting newborn mice with anti-NFG antibodies. The result was a dramatic atrophy of the sensory and sympathetic ganglia. It was concluded that neurons that were in touch with peripheral tissues via their axons transported NFG upstream and survived, whereas those that didn't, degenerated and died.

The discovery of growth factors depended on both chance and convergence. The role of intercellular communication in embryonic development had been discovered in the laboratory of Spemann in Freiburg, Germany, where Hamburger had trained. G. Levi studied development of the nervous system at a more microscopic level. Levi-Montalcini used a modification of Ramon y Cajal's staining technique to distinguish individual neurons. G. Levi also pioneered the use of the *in vitro* culture of tissues and organs. Hamburger had a holistic view of development, Levi-Montalcini a more reductionist and molecular one. Both visions contributed to our understanding of the nervous system as being adaptable, plastic, and not rigidly determined. NFG, was the first of a series of small peptide factors that stimulate cell growth or division, or inhibit programmed cell death. NFG got its name in 1954. Nowadays, searching for it yields 51,295 entries. Growth factors may influence various cell types differently. They are all recognized by receptors on the cell's surface, which transmit a signal that informs a cell of its fate: grow, divide, or even die. Their role in the immune response, in pathological changes, and in therapy has become central to modern medicine. The contribution of a deficit of NFG in Alzheimer's is being actively investigated and discussed.

Life after the Nobel Prize

The 1986 Nobel Prize in Physiology and Medicine was awarded to Levi-Montalcini and Cohen, surprisingly omitting Hamburger. Since 1962, Rita divided her time between St. Louis and Rome, where she directed a center of cellular biology (1969–1979), founded the EBRI Institute of Neurobiology, and kept up her scientific and community activities. In the Senate, her deciding vote ensured the majority for Prodi's government (2006–2008), which earned her insults from right-wingers. She participated in the FAO Action Against Hunger. In 1992 she started a foundation to promote access to education of African women. She wrote eight books. At one point, she declared

Salvador Luria.

Renato Dulbecco.

that her brain worked better at one hundred years of age than at twenty. Rita Levi-Montalcini embodied the intellectual and ethical values that are the absolute opposite to the vulgarity and banality afflicted on Italy by the Berlusconi era.

Claudio Scazzocchio is Visiting Professor in the Department of Microbiology at Imperial College London. This article appeared originally in a shorter form in Brecha, *(Montevideo) January 25, 2013, and is reproduced here, with permission, in translation with a few modifications.*

Footnotes

[1]The first graduate student of Salvador Luria was no other that Jim Watson. The description of Rita Levi-Montalcini meeting the not yet twenty years old Watson, is worth reading. The following is a very free translation from the Italian by CS. The time is "late autumn, 1947," the place, Bloomington, Indiana.

"*At the end of our conversation, Luria introduced me to one he defined as his most gifted student, who has joined his team not very long ago. It was Jim Watson, not yet twenty. He looked adolescent, with a blond tuft of hair, shading his brow, which in time would be replaced by precocious baldness. His engrossed and dreamy appearance, his strut and thinness, reminded me of a harlequin of Picasso's blue period. He took absolutely no interest in my person, and left swiftly with a hurried good-bye. Later on, when the unknown teenager had become the famous Watson, our sporadic encounters were characterised by the same complete indifference to my research and myself. I was aware of his misogyny and his coldness never bothered me.*"

[2]An inkling of Levi's character, as a mentor and as father can be had by reading Rita Levi-Montalcini autobiography, *In Praise of Imperfection* and in *Family Sayings*, one of the books of Giuseppe Levi's daughter, the novelist Natalia Ginzburg. It may be of interest that Natalia Ginzburg's son, Carlo Ginzburg is a major Italian historian.

February 25, 2013

bit.ly/1Llfcz3

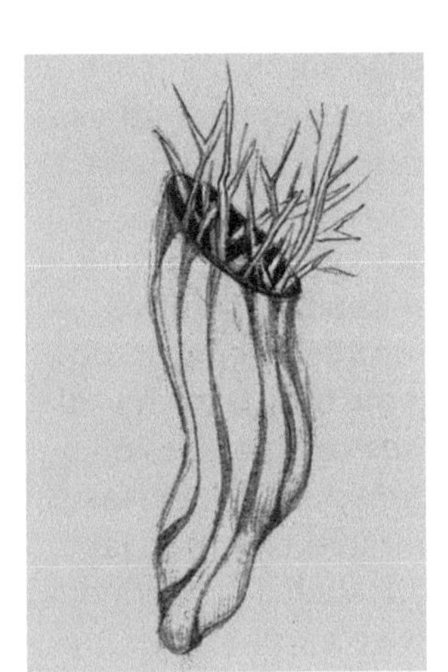

24

A Failed Experiment

by Elio

In 1956 I joined Ole Maaløe's laboratory in Copenhagen for a two-year postdoc. We worked on the connection between the rate of growth of *Salmonella* and its macromolecular composition, arriving at the conclusion that there was indeed a simple linear correlation between the cells' nucleic acid and protein content and how fast they were growing. In trying to interpret this, Ole was influenced by the experiments coming from the labs of quite a few people, showing that the synthesis of many biosynthetic enzymes becomes repressed when the end product of their pathway is added to the medium. If, say, arginine is added to a culture, the enzymes involved in arginine biosynthesis will not be made. If so, in cultures grown in a rich broth, many biosynthetic genes should indeed be silenced. These should include all the operons for amino acid biosynthesis and for other building blocks present in this medium.

Ole Maaløe.

Ole reckoned that if he kept a culture growing steadily

for a long time in rich broth, the cells might shed some of their biosynthetic genes because they would not be needed under these conditions. So he did the following experiment: he inoculated a large flask containing perhaps 3 or 4 liters of a very rich broth. As an aside, the kitchen at the State Serum Institute where Ole worked until he became professor at the University of Copenhagen was known for making exquisitely rich media. The ladies who worked there prepared their own meat infusion using the best Danish veal meat in the market and added to it carefully selected batches of peptone. Their broth allowed *Salmonella* to growth at a superfast 16 minute doubling time (try that with dehydrated commercial media!) So, you can expect that the cultures in such a superrich medium might be super-repressed.

Ole's inoculum was small enough that the culture was still in exponential growth by the time it was time to go home. I'm quite sure that in order to slow down the growth he kept the culture at a temperature lower than the 37° optimum (but don't remember what it was). Before leaving the lab, he inoculated a second flask using a small aliquot from the first flask. The next morning he did the same thing, and he kept up this series of twice-a-day inoculations for (I think) one week. By the end of this time, the culture had been growing exponentially for perhaps 500 generations. He reasoned that this careful protocol was necessary to avoid the genes reawakening in the stationary phase, where they may be needed. Thus, even an incipient entry into stationary phase had to be studiously avoided.

A signpost in a Copenhagen street named after Ole Maaløe, shown here with his grandson.

I don't recall the size of the inoculum he used for each transfer. Note that this is a crucial point because this step stacks the experiment against itself by creating a population bottleneck (a sudden decrease in population size).

At the end of the experiment, Ole plated the cultures on a large number of Petri dishes with agar made with the same broth. He then replica plated a large number of colonies (let me guess a total of 10,000) onto minimal medium agar plates containing glucose as the sole carbon source. He expected to find that some colonies from the broth plates would not grow on the minimal medium. These would represent auxotroph, cells that had lost biosynthetic genes due to some mutation or other, preferably a deletion. To his chagrin, he found no such colonies. In conclusion, he had laid waste to a lot of broth and Petri dishes, but at least got his answer in one week.

I have wondered about this experiment in the intervening years. Of course, many *in vitro* evolution experiments have been carried out since, but best I know, none using this particular protocol. So, over 50 years later, what might be the reason for this failure? I can come up with a few possibilities that I'll list below, but I invite everyone to contribute their thoughts.

1. The experiment was not carried out long enough. It may take longer than 600 generations to accumulate a significant number of mutants in a dispensable function in the sample size he used.
2. The selective advantage of losing a relatively small proportion of genes may be small. This could be tested directly by doing competition experiments using wild type and auxotrophic cultures, which someone has probably carried out already.
3. Biosynthetic genes may have multiple functions, some of which are needed even when the genes are repressed, which sounds a bit mystical. What could they be? Keep in mind that repression is usually not absolute and that some residual expression is likely.

I enlisted the help of three people in the know. Here is what they had to say:

Kevin Foster, Professor of Evolutionary Biology, Department of Zoology, University of Oxford, pointed out that 500 generations is not a long time. He further suggested that "*the minimal medium used gave the bugs glucose and there will be glucose and glycogen* *(which will be converted to glucose by the bugs) in meat broth. So the initial selection would have used the same enzymes as the final selection. To do this experiment again, I would recommend making the metabolic requirements non-overlapping. But it would also have to run for longer AND I would, ahead of time, try to find a metabolic process that is known to be costly to run.*" Furthermore: "*auxotrophic mutants do not show much of a growth advantage if any at all (but we don't work with these much)*."

John Ingraham, Professor emeritus, University of California at Davis, focused on the fact that for auxotrophs to become detectable, they would have to have a growth advantage over their prototrophic parents. "*The growth advantage of an auxotroph would derive from its saving ATP (owing in the case of a deletion to making a tad less DNA, and in all blocking mutations by decreasing flow through an unneeded pathway below the perhaps residual flow escaping feedback inhibition, and making less of the unneeded enzyme than the level set by end-product repression) and probably to a lesser extent saving carbon.*" He then said that "*Ole's medium is most probably one that would offer ATP and carbon excess. (Who knows what set its growth rate, but it supports a blazingly fast perhaps not improvable growth rate.) Instead, Ole should have used a minimal medium with a poor carbon source (for example succinate or lactate) in which ATP and carbon supply limit growth rate. Then end products that are not utilizable as sources of energy and carbon should be added to support growth of corresponding auxotrophs. These would include certain amino acids, uracil, purines, and certain vitamins. Such a medium, I would think, would offer a distinct growth advantage to auxotrophs.*"

Kevin Young, Professor, Department of Microbiology and Immunology, University of Arkansas for Medical Sciences, calculated that the inoculum size must have been too small to contain any mutant that may have arisen. He posited: "*I don't think the experiment failed in the sense of disproving the original hypothesis (that auxotrophs would accumulate when grown in a rich medium). I'm just afraid that the repeated re-inoculation steps continually selected for the most prevalent cells in the culture, which were the un-mutated wild type cells.*"

Elio adds: I agree that this last point is crucial and may suffice to explain the apparent failure. One way to repeat the experiment without this limitation could be to use a turbidostat, a device that allows constant growth at a maximal rate and a fixed population density. No evolutionary bottleneck here. This would overcome the major object of Kevin Y. and could be used to probe into the experimental situations suggested by Kevin F. and John I.

February 27, 2014

bit.ly/1GfyGX3

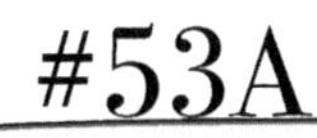

#53A

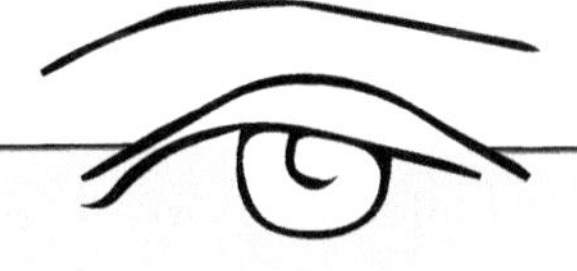

by Elio

In recent years, the use of radioactive isotopes has diminished considerably in microbiological research. What has been lost in the process?

September 28, 2009

bit.ly/1Nl4KIG

25

Microbe Hunters by Paul de Kruif: A Major Force in Microbiological History

by Elio

Chances are good that when you ask senior microbiologists how they got into this science, they will tell you that they were influenced as teenagers by reading *Microbe Hunters*. I was about 14 years old when I did, and I became transfixed by the glorious and intrepid achievements of our forefathers, told in a jaunty and florid style. There was nothing else I wanted to do with myself than to become a Microbe Hunter.

This book was written in 1926, two years before my birth, by the microbiologist Paul de Kruif, whose career was as checkered as it was unusual. He graduated in 1916 from the University of Michigan with a PhD in microbiology and, after World War I, got the job at the Rockefeller Institute. This, however, did not last long because his lifestyle included writing for the general public, which at the time was frowned upon by the establishment. He also hung out with such suspect characters as Carl Sandburg, H.L. Mencken, and Sinclair Lewis, of whom he became a welcome drinking companion. De Kruif became the source of microbiological information for Lewis's best-selling book, *Arrowsmith*. In this book, a somewhat persecuted microbiologist works on developing phage therapy in a lab he had built in his own home.

Microbe Hunters rapidly became a huge success. It was translated into all major languages and inspired two Hollywood movies. It still sells well. But for all its appeal, it received serious criticism. De Kruif wrote in a "jazz style," where bravura at times seems to be more important that the facts. To make the narrative conform to his "voice," de Kruif not only made up dialogue from whole cloth but also took liberties with the facts. So much so that one of the people depicted in the book, the British microbiologist Ronald Ross, threatened a lawsuit. The matter was resolved by omitting the chapter on Ross from the British edition! But such thorny facts do not detract from the impact of the book. As I can attest, if you read it at a susceptible time of life, you will be carried away by the pungency of the tales, the spirited portrayal of the protagonists, and the glory of the achievements.

April 3, 2014

bit.ly/1jR6uzi

by Daniel Portnoy

Why don't mammals make antibiotics of the type so readily made by bacteria and fungi?

March 19, 2009

bit.ly/1Nl4JV2

26

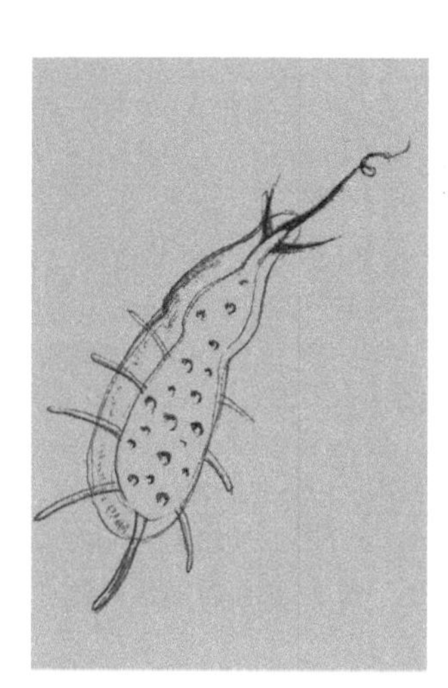

Fecal Transplants in the "Good Old Days"

by Stanley Falkow

I had a conversation with some colleagues last week about "personalized medicine," which has been transformed now into the term "precision medicine." The conversation revolved around what to do about the perceived effects of antibiotic treatment on the microbiota of individuals. How does one treat a patient without disrupting his or her microbiota? Do we create new classes of antimicrobials that target only a precise pathogen? I opined that I thought the day was coming when all individuals might have the microbiota from each anatomic site preserved so that it could be reconstituted after some catastrophic disruption caused by antimicrobial therapy for an infection, transplantation, surgery etc. The topic of fecal transplantation and how successful it has been for the treatment of intractable *Clostridium difficile* infection then came up. Would fecal reconstitution really work?

I answered truthfully that I did not know, but my experience many years ago led me to believe it would. One of the people in this conversation, John Mekalanos, no stranger to stools, asked when I participated in a fecal transplantation study. It occurs to me that my experience in this study might be of interest to, or at least titillate, those who read this blog.

I was a 23-year-old medical technologist (MT, ASCP) in 1957 working as a journeyman bacteriologist in several clinical laboratories in Rhode Island and Massachusetts. It was a time when the *Staphylococcus aureus* 80/81 phage type was raising a specter of uncontrolled hospital infections. It was a time when large doses of antibiotics were administered to patients pre-operatively and continued until they were discharged some days later. Many of these patients reported to their physicians that they suffered from diarrhea, flatulence, and indigestion, and generally felt terrible after their surgery, though the operation was deemed a success. This was before anyone knew about *C. difficile*, of course, but antibiotic-associated diarrhea was known even in those days. One of the internists I knew well came to talk with me. I should point out that in the late 1950s many (most) of the physicians were frequent visitors to the bacteriology laboratory because they wanted to look at the Gram stains and the cultures obtained from their patients. The physician in question, who I will simply call Dr. S, thought, after examining and talking to patients who had not "felt right" after their surgery, that they had suffered from the aftereffects of the antibiotics that had been given them to sterilize their bowel flora before surgery. The feces of many of these patients would yield no growth on blood agar plates and MacConkey agar for days after their surgery. (We didn't do anaerobic cultures in those days, though.) The stools were even odorless. Few stools can make that claim. S thought that their normal flora had been disrupted by the antibiotics. "*Healthy bowels and regularity make a happy patient*," he said.

Dr. S thought it would be prudent to ask patients to bring in a stool specimen when they came to the hospital for their surgery. He said to me, "*Now, Stan, how do we get it back into them?*" We decided that gelatin capsules from the pharmacy might do the trick. The pharmacies at that time still made a good deal of their own formulary. We set up a protocol. Stools were obtained from the patients immediately after admission. I would transfer the stool as quickly as possible into 12 large gelatin capsules. This was a messy and not a precise or enthusiastic process on my part. I would wash the capsules in water and rinse in a dilute solution of mercuric chloride to disinfect the outer surface of the capsule, and then the capsules were rinsed again and put in a small ice cream carton and put in the refrigerator labeled only with the patient's initials.

Upon discharge, Dr. S and one other physician who became a convert to this "protocol" would obtain the capsules from me. They would tell the patient to keep them refrigerated, to take 2 twice a day until they were all consumed. At least Dr. S told them, "*Eat lots of salad.*" This uncontrolled trial continued for some months and, according to the anecdotal reports of Dr. S and his colleague Dr. B, was quite successful in comparison to the patients of other physicians who did not have the benefit of the autogenous fecal sandwich. I don't recall that we ever thought about the ethics of this. It was a time before informed consent. I'm pretty sure, however, that the esthetics of this practice was understood and that the patients in question never knew the contents of the capsules they ingested.

The chief hospital administrator discovered what was up. He confronted me and exclaimed, "*Falkow, is it true you've been feeding the patients s**t!*" He used the Anglo-Saxon phrase for feces. I responded: "*Yes, I had been a participant in a clinical study that involved the patients ingesting their own feces.*" "*You're fired!*" was the reply, although Dr. S came to my rescue. I was rehired two days later. Thus, the "experiment" came to an abrupt end. I left in June, 1958, to study for my PhD with C. A. Stuart and Seymour Lederberg at Brown University.

Now I am not going to claim that I knew that feeding patients their own feces after intense antibiotic therapy would be beneficial. Dr. S was sure it was the case based on his years of clinical experience. I understood the point that the indigenous flora was important. I had examined hundreds of stool specimens from sick and well people for too long. I routinely Gram-stained fecal samples and examined them in a wet mount. I didn't have deep sequencing but I could discern differences in the flora of individuals. One fecal flora did not reflect all. I understood this even better after I met René Dubos and Russell Schaedler at the Armed Forces Epidemiology Board in the early 1960s and even more when I read the wonderful book by Theodor Rosebury, *Life on Man*.

My experience presaged the current excitement and exciting information that has deluged us in the past few years about the wonders of the human microbiota. The understanding and the appreciation for the sanctity of the "normal flora," however, is not a new thing. Fecal transplants, of a sort, were practiced some 50 years ago because of the recognized untoward effects of antibiotic therapy. Mekalanos, after hearing this story, said: "*It once was 'Eat s**t and die!' Maybe now it will be 'Eat s**t and live!'*"

This experience also reminds me of something I have learned over the years. Experiences that occur during experiments or facts learned in a seminar or read in a paper have a way of reappearing, often decades later, with new meaning.

Stanley Falkow is Robert W. and Vivian K. Cahill Professor in Cancer Research, Emeritus, and Professor Emeritus of Microbiology and Immunology, Stanford University School of Medicine. He is also a past president of the ASM, where he has been a member for 61 years.

May 13, 2013

bit.ly/1klI29C

#57

by Frank Harold

Does anyone know of a solid example of a biological membrane that arises *de novo*, rather than by the extension of a preexisting membrane?

January 7, 2010

bit.ly/1M3pQxR

27

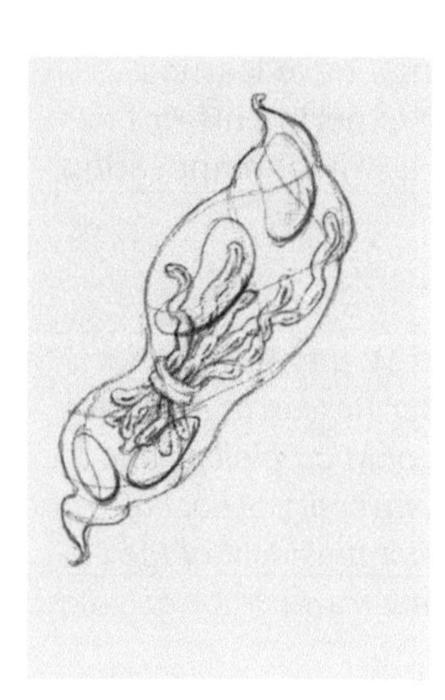

Requiem for a Machine

by Elio

Here's a challenge for present-day systems biologists. Say you wanted to find out how many ribosomes are present in cells growing under different conditions. How would you do it? You might think of using quantitative PCR to measure the amount of rRNA inside the cell. However, you could end up with an overestimate because some of the rRNA may not yet be assembled in mature ribosomes. For instance, add chloramphenicol, a protein synthesis-inhibiting antibiotic, to a culture and you will measure increases in the cellular levels of rRNA due to the accumulation of non-functional, immature ribosomes. Perhaps, then, you would roll up your sleeves and run a sucrose gradient to separate the mature ribosomes from their precursors and immature forms by size fractionation, but this is a labor intensive method that not many people like to do these days. We'll agree that making such measurements may not be as easy as it sounds. So let me reminisce about how we carried this out in the old days. All it took was a fancy apparatus and some chutzpah.

From about 1950 to 1970 or so, the most conspicuous device in microbiology and biochemistry departments, except for the electron microscope, was the analytical ultracentrifuge. Made by Beckman Instruments, it was known as the Spinco Model E (not to be confused with the Jaguar type E), and it was the pride of research departments. It was an imposing machine indeed, about the size of two refrigerators. It was originally developed for applications requiring determination of the purity, molecular weight, and molecular interactions of proteins and nucleic acids. Meselson and Stahl used a Model E for their classic experiment that demonstrated the semiconservative replication of DNA. In time, various kinds of gel electrophoresis machines appeared on the scene that allowed measurement of many of the same properties of macromolecules. The Model E now sat quietly idle.

In the late 1950s, we used my department's prized Model E to determine the ribosome concentration of bacteria growing at different growth rates. (This followed earlier studies where we had measured total RNA as a function of growth rate.) How did we do this? We broke bacteria apart and— this is where gall came in— introduced the entire lysate straight into the cell of the centrifuge. One usually only offered purified preparations to the Model E, and here we were sullying it with an utterly uncharacterized gemisch. But we did just that, thanks to the people who taught us this technique—the physicists-turned-biologists at the Department of Terrestrial Magnetism of the Carnegie Institution of Washington (more about them later).

After we loaded our sample and turned the machine on, the "debris fraction" containing cell envelope fragments sedimented to the bottom of the cell as the rotor came up to speed. The next heaviest components, thus the next to spin down, were the ribosomes. We could now easily and quite precisely determine their relative amounts, as the special (Schlieren) optics of the Model E yielded a Gaussian curve whose area was proportional to the amount of ribosomes. But how to get from this number to ribosome *concentration*? What we did is to keep the machine running until the next large fraction sedimented into view, this fraction being the soluble proteins of the cells. Amazingly, this mixture of thousands of proteins also followed a Gaussian curve, suggesting the odd fact that the molecular weights of these proteins were nearly normally distributed. In any event, we now could determine the ratio of ribosomes to soluble protein, a reasonable measure of ribosome concentration. This method was simple, accurate, and not terribly time consuming, once you got used to all the buttons and dials. But it wasn't all that long afterwards when the once useful behemoth of a machine became obsolete and, with considerable sadness, we shipped it to the dump. Analytical ultracentrifugation is enjoying a rebirth, thanks to a modern line of analytical ultracentrifuges called Optima X produced, as was the Model E, by what now goes by the name of Beckman-Coulter.

A Word about the Department of Terrestrial Magnetism of the Carnegie Institution of Washington

Arguably, the pioneering work of a group of ex-physicists at this institution stands as a forerunner of modern-day systems biology. I am not sure why they worked in a department with a name so distant from biology, but I remember that at least some of them had worked together during World War II to develop the "proximity fuze (sic)," a device that triggered a bomb to explode above the intended target.

For reasons that escape me, this group decided to work on many of the biochemical steps involved in biosynthesis by *E. coli*. This they did with a combination of insight and simplicity. Basically, they grew the organism in a defined minimal medium, added one of a number of radiolabeled compounds that would be incorporated into specific metabolites, then fractionated the cells and measured the amount of radioactivity in various cellular components. Note that, at the time, the field had been dominated by biochemical studies that almost exclusively used cell-free extracts prepared from ground-up bacteria. Their novel experiments, carried out with unusual precision, revealed not just pathways but also kinetic parameters. By looking at everything in the cells, not just some predefined string of biochemicals, they unearthed some unexpected (and, I suppose, forgotten) facts. One that sticks in my mind is that about 20% of *E. coli*'s proteins turned out to be soluble in ethanol! I would never have guessed.

Their work was only occasionally published in the research journals. Rather, they presented their work in reports to the Carnegie Institution. These culminated in a book entitled *Studies of Biosynthesis in* Escherichia coli, which went through two printings (1955 and 1957). It was authored by R. B. Roberts, D.B. Cowie, P. H. Abelson, E. T. Bolton, and R. J. Britten. An abstract can be found here (bit.ly/1S8qFo1), and the book can still be obtained from used book dealers. It's a genuine classic and very much worth reading.

August 23, 2012

bit.ly/1Gnov2r

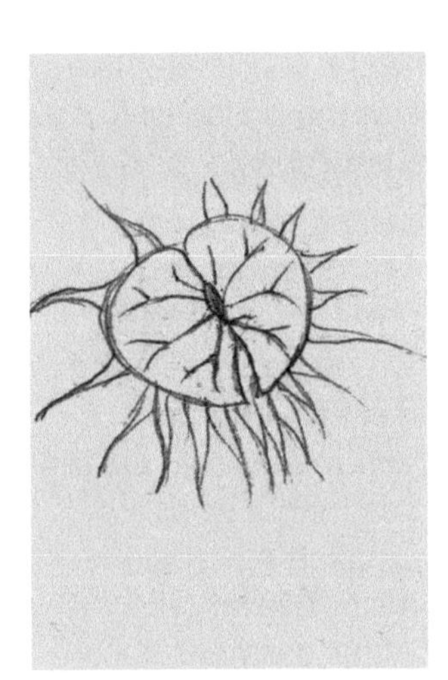

28

Mad Dogs and Microbiologists

by William C. Summers

Jean-Baptiste watched the powerful dog with unsteady gait approach and then attack a group of six of his friends. He picked up his whip and rushed to meet the animal, only to be savagely bitten on his left hand. In fierce hand-to-hand combat, Jean-Baptiste finally managed to throw the animal to the ground, pinning him to the ground under his knee. With his right hand he forced the dog's jaws apart, sustaining new bites, then used his whip to tie the muzzle of his enemy and finally beat the beast to death with one of his wooden shoes.

Thus it was that Louis Pasteur recounted to his colleagues at the Académie des Sciences in Paris how his heroic patient, Jean-Baptiste Jupille, a 15-year-old shepherd boy from the Jura, came to be exposed to rabies on October 14, 1885. Pasteur was reporting on the success of his treatment of his first rabies patient, Joseph Meister, who was still alive and well more than three months after having been severely bitten by a rabid dog. Meister's survival was considered something of a miracle, because rabies was, and still is, considered a lethal disease in the absence of effective treatment. Young Jupille was treated as Pasteur's second rabies patient, and he, too, survived this heretofore universally fatal disease.

Still, Pasteur had been hesitant to treat Jean-Baptiste because the disease had such a head start on him. From his experimental work on dogs and rabbits, Pasteur and his protégé, Emile Roux, knew that their new treatment was most effective before or very soon after inoculation of the infectious agent. The longer the interval between inoculation and the start of treatment, the less likely the cure. This very early conclusion remained one of the unsolved problems in rabies treatment, at least until very recently.

Since ancient times rabies (known also as "hydrophobia" or "la rage") has been a dread disease with virtually 100% mortality. The rare survival was usually attributed to supernatural intervention. Most often invoked in miraculous cures was Saint Hubert, the first bishop of Liege in Belgium and the patron saint of hunters, hence associated with dogs. The standard legend relates that during his investiture as bishop, one of the required vestments was missing. At the last moment, it was miraculously supplied when Saint Peter appeared with a stole woven on the looms of heaven and embroidered by the Virgin Mary. Threads from this garment, inserted into a small incision in the forehead of the rabies victim, became a sure cure for rabies—provided some additional rituals were observed. Another legend describes cauterization of the wound site from the bite of a rabid animal with a heated "key of St. Hubert" in the form of a metal cross. Even now, on St. Hubert's feast day, November 3rd, in Ghent local pastries known as mastellen are taken to Mass for blessing, thus becoming remedies for rabies. It is reported that in Flanders there are open air Masses where

The Vision of Saint Hubert by Jan Brueghel the Elder and Peter Paul Rubens. Early 17th century.

Source: Corbis images, Rights Managed

hunting dogs and other animals are given "The Bread of Saint Hubert" as a sort of spiritual rabies shot.

The standard treatment for rabies is a direct descendant of Pasteur's original concept: injection of attenuated virus before the infection reaches the central nervous system and before it does substantial damage. In this way, one can induce sufficient immunity to destroy the advancing infection. For immunogen, Pasteur employed a preparation of dried spinal cord material from infected rabbits. Drying for varying lengths of time provided virus preparations of varying virulence. He injected the exposed patients with increasingly virulent samples as their induced immunity increased. This procedure was based on Pasteur's theory of the mechanism of immunity. Although his theory has been discredited, he nevertheless devised a regimen that does, indeed, provide effective treatment for rabies victims. While the source of the attenuated virus for immunization has moved on from desiccated rabbit spinal cords to duck eggs, then to cultured cells, the approach has remained the same: induce antiviral immunity before the virus can damage the brain. Thus, to be effective, treatment must begin soon after inoculation.

Both Pasteur's dilemma and the need for miracles, however, may soon be relegated to the dustbin of medical history. A recent report (1.usa.gov/1RdjpX6) in *PNAS* by Bernhard Dietzchold and his colleagues at the Thomas Jefferson University in Philadelphia suggests that a new approach to rabies vaccines may make it possible to treat rabies infections effectively even after some time has elapsed. Drawing on new understanding of the pathogenesis of rabies, including its replication and its expression of the relevant immunity-inducing antigens, these workers designed a highly attenuated—but highly effective—vaccine strain of rabies that shows promise as a significantly improved post-exposure treatment for rabies.

There have been two major obstacles to rabies vaccines: relatively weak immunogenicity and the problem of attacking infections across the mysterious blood-brain barrier. The new work discussed here seems to overcome both of these obstacles. First, in the vaccine strain, the gene for the major antigen required to induce antiviral immunity, the virion surface glycoprotein (G), has been triplicated. Apparently simply because of the increased gene dosage, the G protein is substantially over-expressed. Because

G is also involved in the neuropathogenicity of the rabies virus, two mutations were constructed in the G gene that destroy its pathogenicity, but not its immunogenicity. The authors call this mutant form GAS. Interestingly, this avirulent GAS mutant is dominant over the pathogenic strain, thus ensuring that the "triploid" GAS virus has a very low probability of reversion to the virulent wild type virus. The use of two mutations in G further reduces the probability of reversion to wild type. These predictions were confirmed by demonstrating that the GAS strain is non-virulent even when directly inoculated into the central nervous system of immune-compromised mice.

The second problem, that of immune access to the privileged nervous system, the site of the rabies infection, seems to be overcome by this vaccine strain, as well. Paradoxically, one of the features of the rabies virus that is important in its pathogenicity is its low level of replication. This "stealth strategy" seems to allow it to infect neurons without breaking down the barriers to entry of antibodies. The effective vaccine strains, including the new triple GAS strain, replicate much more rapidly than the wild type virus and, by some mechanism that is still unclear, enable the induced, antiviral antibodies to penetrate the nervous system.

When experimentally testing for post-exposure prophylaxis in a mouse model, the GAS vaccine effectively prevented any signs of infection when administered a short time after inoculation, and reduced deaths and/or paralysis when given at longer times after inoculation. Since it is known that the distance between the site of inoculation into peripheral nerves and the central nervous system is a crucial factor in determining the "window of opportunity" for effective post-exposure prophylaxis, it can be expected that the short "window" in mice will translate into longer time intervals in humans simply based on host size and the distance the virus must travel.

Nearly 125 years after the momentous introduction of the rabies vaccine, we may finally be able to close the circle. The new vaccine, like Pasteur's, also involves an attenuated strain. One would guess that Pasteur would have approved.

Happy World Rabies Day (28 September 2009)

William C. Summers is Professor of Therapeutic Radiology, of the History of Medicine, and of Molecular Biophysics and Biochemistry, as well as Lecturer in History at the Yale University School of Medicine.

October 26, 2009

bit.ly/1GfyQ0s

#58

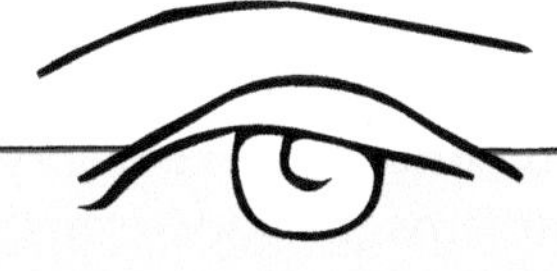

by Elio

What if all phages on this planet went on strike and refused to have their genes expressed?

February 4, 2010

bit.ly/1klquKR

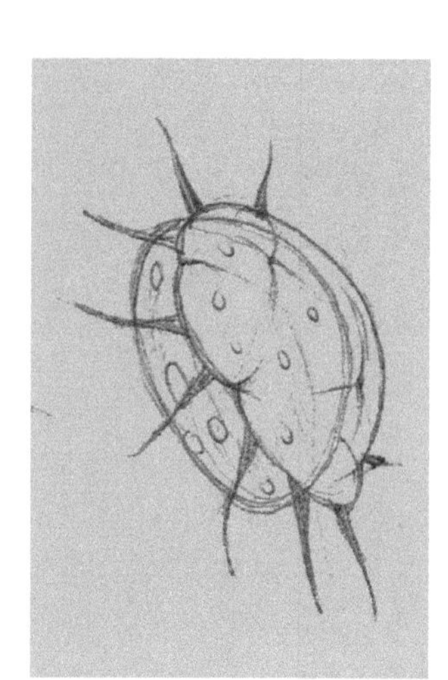

29

Joshua Lederberg and the "Cellularity" of Bacteria

by Elio

Joshua Lederberg died recently. His is a revered name among our older readers, and we assume that many younger ones also know of him. But just in case, let us mention some of what made him a giant in modern biology. He is best known for finding that bacteria have sex, that is that they can undergo cell-to-cell conjugation. He then discovered (with Norton Zinder, a graduate student) that genes can also be exchanged via phages: transduction. Thus, Lederberg discovered two of the three major mechanisms for gene transfer in bacteria. The third, transformation by naked DNA, had already been discovered, or else—who knows?—he might have found that one, too. He then entered the fray related to the origin of mutations. For this purpose, he invented, together with his wife, Esther, the replica plating technique, which allowed him to demonstrate that mutations were spontaneous and random, not induced by something in the environment. He did this by comparing a mutant colony identified on a replica plate with the corresponding colony on the master plate, and showing that they both had the same properties. That ended the argument, at least for the time.

These accomplishments place Lederberg among the great biologists of all times. However, Lederberg did much more.

After becoming one of the youngest Nobelists in history, he went on to dedicate himself to matters such as exobiology (a term he coined), public policy about infectious diseases, and bioterrorism.

Let me expand on why his contribution transcends the phenomena he discovered. By the mid-1940s, it was becoming evident that bacteria were cells. But were they cells like all other cells, or were they something different? Many biologists were still reluctant to admit bacteria into the realm of "regular" cellular beings. Lederberg believed that sorting this out was of central importance. This is revealed by the interest he took in bacterial cytology, including his finding that penicillin induces the formation of spheroplasts. But it was his discovery that bacteria could mate that swept away the remaining hesitancies. Suddenly, the door was opened for bacteria to take their place as cells like all others—no longer some biological oddity. Lederberg's discovery was most decisive in bringing about this new awareness.

I would also like to share with you this response received from Robert Murray, one of the founders of bacterial cytology.

> "*You are right in saying that Lederberg really had in mind that bacteria were real cells. I know this because both C. F. Robinow and I spent time with him before and while he was at Wisconsin; he was very much interested in the cellularness of bacteria. I am pretty sure that a good part of the stimulus came, as it did for me, from reading René Dubos's book,* The Bacterial Cell, *and thinking about the micrographs in the addendum in that book by Robinow. Lederberg implies that indebtedness in the foreword he wrote for a compendium of classic papers*, Microbiology: a Centenary Perspective, (*unfortunately not all I would have put in*) *edited by Joklik et al. and put out by ASM Press in 1999. Look at his second page where he states: 'A major conceptual turning point was the publication of René Dubos's* The Bacterial Cell *in 1945 just in the midst of the wave of discovery of spontaneous mutation, of genetic transformation, of (conjugal) genetic recombination — i.e., sex — in bacteria, and soon after virus-mediated transduction.' My, that was discovery plus something special that came to a mind prepared to think hard about experiments that told one something. I suppose one could add the influence of Cold Spring Harbor and a few other good minds of the time. Nobody seems to really remember Beadle and Tatum, who got Josh on his way.*

"Lederberg did something for plasmids, too, and I remember him setting me straight on bacteriocins. He mattered to me. I talked to the graduate students this morning about him as the preliminary to a weekly seminar—so also thank you for stimulating that bit of teaching. The three virologists (organizers of the event and the speaker) had not heard of Lederberg. So your blog is worthwhile. The students wanted a date or two as a way of organizing their historical thoughts. I asked the class who had 'already discovered' transformation by naked DNA and nobody, including faculty present, could answer! The years 1945 to 1953 put a lot of the basis of modern biology into focus."

February 7, 2008

bit.ly/1RTnUqC

#62

by Elio

What if all prokaryotic plasmids on this planet went on strike and refused to have their genes expressed?

May 27, 2010

bit.ly/1QNxWss

PART 3

Small Wonders

The blog on which this book is based has been geared towards stories from the microbial world—like these—that are unexpected, fascinating, and often neglected.

30

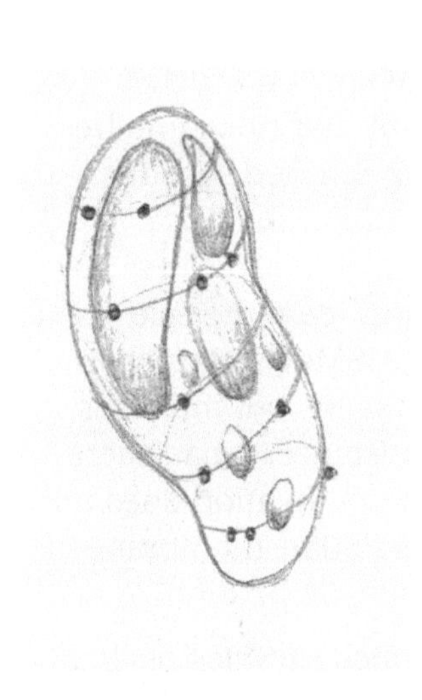

Where Art Thou, O Nucleoid?

by Conrad Woldringh

In a post on May 12, 2008, Elio asked where ribosomes are located in bacterial cells. According to a 2006 paper by Ortiz et al. (1.usa.gov/1NzRno7), "*the ribosomes are evenly distributed throughout the cytosol*" and "*no distinct nucleoid territory was observed.*" This prompts one to also ask, "*Where art thou, O nucleoid?*" Do their observations indeed indicate that, in the small cells of *Spiroplasma melliferum* (Figure E; volume ~0.02 μm^3), the ribosomes (number ~1000) and DNA (genome size 1.46 Mbp) co-mingle and that a phase separation between cytoplasm and nucleoid is absent? In larger, fast-growing *E. coli* cells (Figure B; volume 1–3 μm^3), the nucleoid is clearly visible in living cells as a low-density compartment, as was already documented by Mason and Powelson in 1956. However, distinct nucleoids are difficult to see in smaller, slow-growing *E. coli* cells. In the small *Caulobacter*, nucleoids have even been reported to be absent (R.B. Jensen, 2006) (1.usa.gov/1jQ96ht). The existence of a discrete DNA phase was calculated by Odijk (1998) (1.usa.gov/1RdvvzJ), taking into account the excluded volume interactions between DNA and the soluble proteins as present in small *E. coli* cells (Figure A; volume 0.46 μm^3; genome size 4.6 Mbp). To me it seems unlikely that the physical laws that predict the visible phase separation in the larger cells would not hold for the smaller cells, as well.

Electron microscopy of thin sections of small *E. coli* or *Caulobacter* cells fixed with osmium tetroxide indeed shows a clear phase separation between nucleoid and cytoplasm. An artifact? Maybe, since the same cells fixed with glutaraldehyde show no such phase separation. It is therefore highly relevant that ribosome-free regions with a different texture can be seen in vitrified sections of *Deinococcus radiodurans* (Eltsov & Dubochet, 2005) (1.usa.gov/1RdvNq5) and in cryoelectron tomograms of *Bdellovibrio bacteriovorus* (Figure D; cell volume ~0.08 μm^3; genome size 3.78 Mbp; Borgnia et al., 2008) (1.usa.gov/1GHSMsV). These two techniques are thought to be less prone to artifacts, but this does not mean that artifacts are not possible. For instance, it has been noted (Bayer, 1991) (1.usa.gov/1XxVXI4) that cryofixation does not preserve the adhesion sites between inner and outer membrane of plasmolyzed *E. coli* cells as obtained after aldehyde fixation.

Could the absence of the nucleoid in some vitrified samples, such as those shown in Ortiz's paper, be ascribed to the complex process of freezing? Bacterial DNA is present in the cell in the form of a stiff,

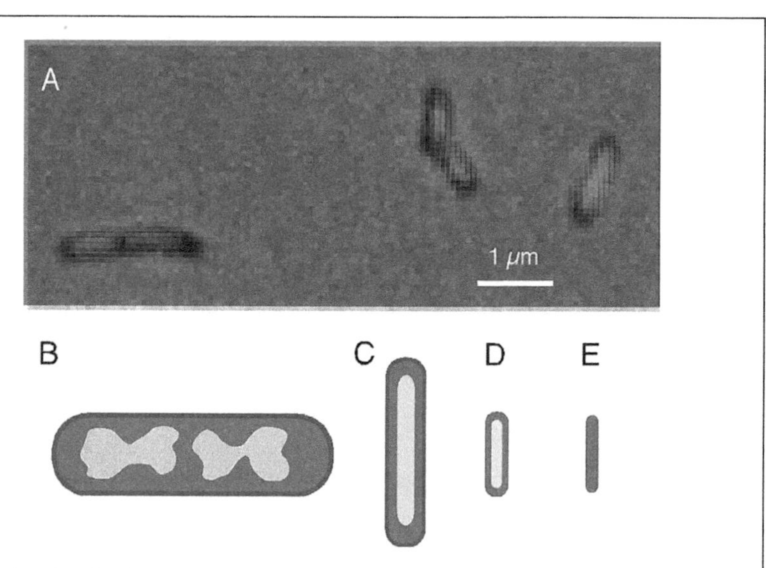

Cell and nucleoid volume of different bacteria drawn to the same scale. (**A**) Living cells of *E. coli* B/r, grown in alanine-medium (doubling time 150 min) and stained with DAPI. Merged phase-contrast and fluorescent images. Figures **B** through **E** are schematic drawings; light areas in **B**, **C**, and **D** represent hypothetical nucleoid volumes. (**B**) Schematic drawing of a fast-growing *E. coli* cell (doubling time 20 min) containing about 4 chromosome equivalents. (**C**) *E. coli* cell grown in alanine medium (doubling time 150 min) containing about 1 chromosome equivalent. (**D**) *B. bacteriovorus*. Dimensions from Borgnia et al., 2008. (**E**) *S. melliferum*. Dimensions from Ortiz et al., 2006. In **D** and **E**, these non-cylindrical organisms are represented as cylinders for ease of comparison.

Credit: Conrad Woldringh.

branched supercoil that, upon osmotic shock, “explodes” like a spring out of a spheroplast. In the intact cell this supercoil is compacted by excluded volume interactions with the abundant cytoplasmic proteins. So, what happens during freezing, a process that of necessity must begin in the extracellular space? Could there be a stage in the process when the motion of the proteins has been slowed, but the DNA could still “spring”? Might this allow them to mix?

Conrad Woldringh is a member of the Swammerdam Institute for Life Sciences at the University of Amsterdam, The Netherlands.

October 13, 2008

bit.ly/1GnoxY5

#66

by Elio

What if someone found an organism whose genes assignation is 1/3 bacterial, 1/3 archaeal, and 1/3 unknown?

September 23, 2010

bit.ly/1Nl4MAf

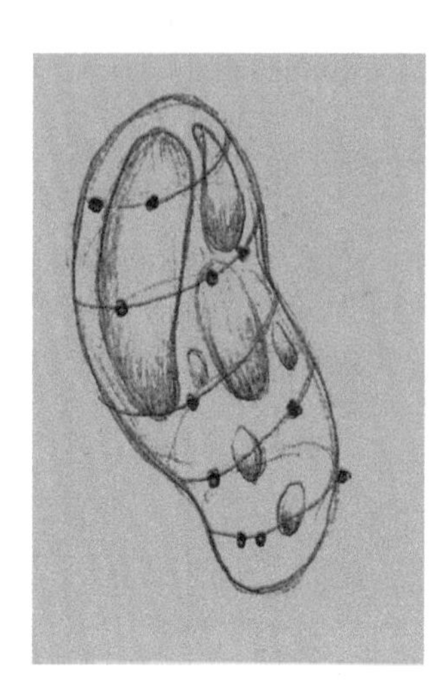

31

Teaching *E. coli* to Endocytose

by Elio

What if I told you that engineering a single protein into *E. coli* is sufficient to make it fill up with membrane-bound vesicles? Would you send me to the couch or to a padded cell? Not so fast, as this is precisely what a group of sixteen investigators from three continents have recently shown by expressing caveolin-1 in *E. coli*. Caveolin-1 is the main protein involved in the formation of mammalian membrane structures called caveolae. The consequences were amazing.

So just what are caveolae? As their name, Latin for *small caves*, implies, they are flask-shaped pits in the cell membrane. Here considerable business is transacted, including the formation of endocytic vesicles and some aspects of cell signaling and lipid metabolism. They are the sites where some bacteria and viruses first attach prior to entering the cell. No wonder this sets the heart of animal cell biologists racing.

The *E. coli* used in these experiments performed spectacularly, beyond any researcher's dreams. They made bilayer-bound, caveolae-like structures that match in key ways the caveolae in animal cells. In fact, they made so many that within a few hours the cells became utterly filled with these vesicles—wall-to-wall caveolae. In our correspondence, the lead author let me know that the gene construct consisted of just the caveolin-1 gene (plus some fusion proteins for ease of manipulation). Thus, this protein alone plus some of *E. coli*'s own lipids is sufficient to induce vesicle formation in *E. coli*. The early stages of formation of these "bacterial caveolae" correspond well to what is seen in animal cells: the cytoplasmic membrane bulges inward and the bulges eventually pinch off to form independent intracellular vesicles. All this appears to happen without further help. When cloned, neither of the two other caveolin proteins do that.

The experiments were straightforward. The researchers introduced the caveolin-1 gene (or those genes for other caveolins in further studies) into a standard plasmid expression vector and transformed it into *E. coli*. To facilitate purification of the protein, this gene had been fused to the one for the maltose-binding protein. The fusion protein was expressed constitutively from the strong T7 RNA polymerase promoter. Cultures of the construct grew vigorously to high overnight cell densities, the same as cells carrying an empty plasmid. By this time the cells were literally filled with the vesicles. Other than that, little is known as yet about how this affects the bacteria.

The caveolae were purified by treating the bacteria with lysozyme and sonication (being small, the caveolae are likely to be more resistant to sonication than the cell membranes), and their proteins were analyzed after affinity chromatography. The researchers recovered abundant caveolin-1, whereas the amount of regular membrane proteins was below the level of detection. In other words, to repeat myself, the vesicles are made from one membrane protein only—caveolin-1 (although periplasmic constituents ought to be found within). Immunogold staining located caveolin-1 on the surface of the purified vesicles.

Looking more closely, a tomographic reconstruction showed that the recombinant caveolae contained some 160 molecules of caveolin-1, similar to the number in the mammalian caveolae. For those interested in the organization of the protein in the structures, this paper provides considerable molecular detail. As to the lipid components of these membranes, a detailed mass spec analysis, including the length of their fatty acids, showed significant differences in the lipids of cells before and after induction of caveolae formation. The caveolae contain more lysophospholipids with fewer shorter chain fatty acids and less phosphatidylethanolamine—changes known to facilitate greater membrane curvature. There is quite a literature on caveolar organization, and this paper contributes interesting possibilities by the use of an easier-to-manipulate experimental system. This work with *E. coli* provides a manipulable model system that may make it easier to address many of the questions that remain unanswered. Although there is already quite a body of literature on caveolar organization, many questions remain unanswered.

Next, these researchers asked if it's possible to introduce foreign constituents into *E. coli*'s caveolae. After incubating the constructs with fluorescent membrane-staining dyes that can cross the outer but not the inner membrane of Gram-negative cells, they saw lots of fluorescence within the caveolae. So yes, the caveolae contain material from the periplasm. And that material can include proteins, as demonstrated by recovering from the vesicles an enzymatically active horseradish peroxidase that they delivered into the periplasm by disrupting the outer membrane. One could call this process endocytosis-like. Something similar has been reported only once in bacteria, to wit, in the somewhat enigmatic Planctomycetes.

Considering what we know about bacteria, this filling up of bacterial cells with extraneous vesicles is simply startling. I know of no precedent. What this tells us about the plasticity of bacterial cells remains to be seen, but it sounds like there is quite a message here. Few details on the bacteria themselves are available as yet, leaving the door wide open for microbiologists to get involved. Some surely will. Questions arising include the following: For how long can *E. coli* do this and survive? Which, if any, cellular processes does caveolae formation interfere with? What accounts for such a copious production of a single protein (to the tune of Mussorgsky's *Sorcerer's Apprentice*)? And how are the synthesis of the protein and specific lipids connected? Time-course experiments monitoring gene expression, protein synthesis, and lipid production following induction could help explain how *E. coli* manages this prolific production of caveolin constituents. The list of questions goes on and on. You are invited to contribute to it.

That this system may be of use in biotechnology is also exciting. Consider the possibilities for drug delivery offered by encapsulation of small molecules and proteins introduced into the periplasm. In another proof-of-principle step, these researchers showed that caveolae containing GFP and an immunoglobulin-binding domain from staph Protein A not only became bound to human breast

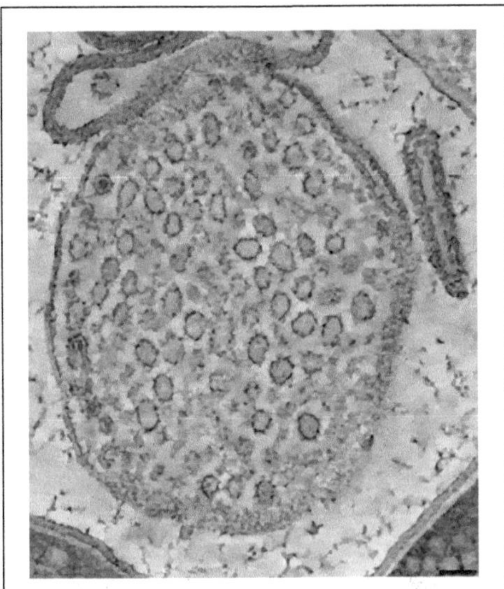

Source:http://www.sciencedirect.com/science/article/pii/S0092867412008860

cancer cells carrying an antibody on their surface, but appeared to become internalized. Being that the caveolin-1 protein can be tampered with significantly without loss of function, one can envisage making fusion proteins targeted to specific cells. It appears that this could become a formidable tool for cancer chemotherapy and for other conditions that may be ameliorated by the targeting of specific proteins. Sounds relatively simple and straightforward. And keep in mind that the caveolin-1 molecule is of human origin. So, will this prove to be a therapeutic breakthrough? Time will tell. But this is certain: our perception of what can be fitted into a bacterial cell has changed. If indeed *E. coli* is only one protein away from endocytosis, perhaps this work will compel us to cross endocytosis off the list of physiological traits that are exclusive attributes of the eukaryotic cell plan.

October 1, 2012

bit.ly/1PDae3J

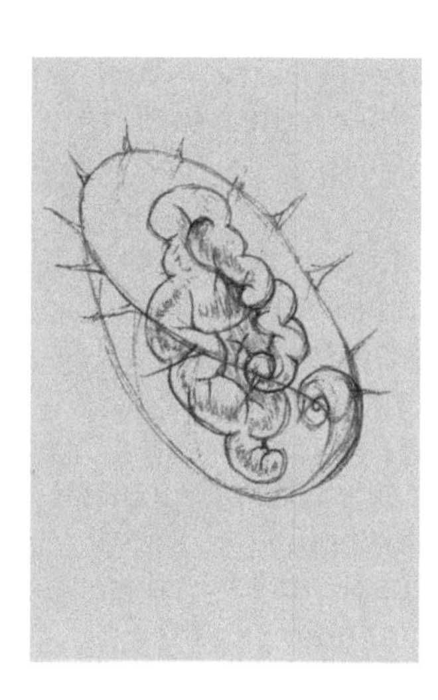

32

The Awesomest Thing in Biology

by Elio

So says a blogger by the name of Psi Wavefunction regarding an eye-like structure of some protists called the ocelloid. She wrote this in a now-defunct blog eponymously called The Ocelloid (bit.ly/1H95b40). I couldn't agree with her more. The ocelloid is the structural and functional equivalent of the sophisticated eye of metazoans—and occurs in an unicellular organism! If you can think of anything that is "awesomer" than that, please let me know.

In the parade of startling facts of biology, one that stands out is the variety of morphologically complex structures made by the protists. Some have a mouth, a gullet, a bladder-like contractile vacuole, an anus, one of a variety of organelles of locomotion including some that look like legs, and yet other complex structures (what, no brain?). This begs the question: why did they not go the multicellular route? What more can unicellular organisms achieve?

But even in this wonderland, the ocelloid takes the cake (even though my spellchecker keeps changing it into "ocelot"). Being a miniature likeness of the vertebrate eye, it attains extremes of biological intricacy. Darwin considered the eye to be hugely complex ("... *of extreme perfection and complication*"), although not irreducibly so. He wisely noted: "*I may remark that, as some of the lowest organisms, in which nerves cannot be detected, are capable of perceiving light, it does not seem impossible that certain sensitive elements in their sarcode should become aggregated and developed into nerves, endowed with this special sensibility.*" ("Sarcode" is an old term for the protist cytoplasm.) The repeated evolution of the eye from a primitive light-sensing device has been convincingly discussed. What is remarkable about the ocelloid is that it represents a colossal jump from the simple photoreceptive layer one finds in many unicellular organisms to a remarkably intricate structure—still within a unicellular organism. The ocelloid is found in some dinoflagellates only and is indeed a unique structure in nature.

The ocelloid's anatomy tells all. It is a roundish body about 20 μm in diameter (roughly 1/5 to 1/10 the length of the cell that bears it), endowed with a cornea-like cover, a prominent lens (called the hyalosome) bounded by iris-like constriction rings, and a complex retina-like body that is concave in shape. You will agree the analogy to a metazoan eye is startling, even if its diameter is nearly 1000 times less than that of the human eye. It makes you wonder if these organisms may develop cataracts in old age!

A new paper from labs in Japan, Switzerland, Taiwan, and Saudi Arabia goes into detail regarding the structure, function, and evolutionary origin of the ocelloid. The retinal body gets particular attention. This "organ" consists of an array of parallel lamellae that are ~40 nm thick in the light and ~50 nm in the dark. The structure resembles the thylakoids, the photosynthetic, stacked membranous compartments of chloroplasts and cyanobacteria. The number of lamellae increases in the dark, as does the surface area of the retinal body itself. Thus, the ocelloid responds to light conditions although, oddly perhaps, the cells that bear it are not photosynthetic and lack chloroplasts. It remains to be seen whether the ancestor of the ocelloid-bearing dinoflagellate was a photosynthetic microbe that eventually donated its chloroplast to become the retinal body.

In many organisms, the cytoskeleton is involved in changing the conformation of photoreceptors in response to the light. Here, actin is found only in the retinal body and changes in its morphology are inhibited by cytochalasin B, an inhibitor of actin polymerization, which suggests that actin is involved in the response to light. Actin is not found in the lens, which is thus unlikely to change shape. To complete the picture, does the ocelloid contain light-sensitive pigments? Indeed it does, in the form of rhodopsin. By *in situ* hybridization with probes to its mRNA, the authors found a rhodopsin-like gene in the retinal body only, which suggests that rhodopsin-like sequences are translated at this site.

Recapping, the ocelloid has the structural, and likely also the functional, properties of an eye. Notably, it has a light receiving structure and, in front of it, a refractive lens. As shown by micro-optometry measurements, the lens facilitates light perception in the dim light habitats of these dinoflagellates by concentrating the light on the receptor surface. Just why they like such light conditions is not quite clear.

Where does the ocelloid come from? Like most respectable organelles, it contains DNA (which, best I can tell, still awaits sequencing). Other evidence for this? The ocelloid is surrounded by a double set of membranes, can be seen dividing, and its retinal body has a thylakoid-like structure likely derived from a chloroplast. The rhodopsin gene appears to be of bacterial origin, hinting that it was acquired by horizontal transmission, possibly from a bacterial ancestor. It is thought that the ocelloid increased in complexity

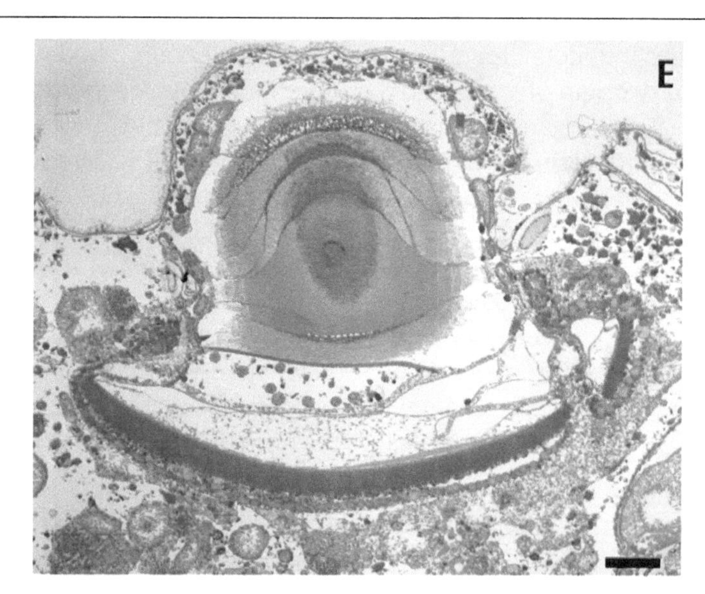

TEM of the dividing ocelloid of *Erythropsidinium* spp. shown in longitudinal section. Bar: 2μm.

Source: **Hayakawa S, Takaku Y, Hwang JS, Horiguchi T, Suga H, Gehring W, Ikeo K, Gojobori T.** 2015. Function and Evolutionary Origin of Unicellular Camera-Type Eye Structure. *PLoS ONE*. 10(3):e0118415. doi: 10.1371/journal.pone.0118415.

during the evolution of these dinoflagellates. For studies of the molecular phylogeny of this group, see here (bit.ly/1RdjoCM) and here (1.usa.gov/1WgMqI3).

Why have these dinoflagellates gone to all this trouble? What do they want to "see"? The answer is not clear. This group of microalgae is not photosynthetic, so they are not looking for light to support an autotrophic existence. Like other protists, they have embarked on extremely sophisticated evolutionary ventures. This particular group, the Warnowiids, have other extraordinary structures besides the ocelloid, structures such as a "piston"—a long posterior "tentacle" that rapidly contracts and expands, presumably for cell motility—and nematocysts—ejectable barb-like structures used by jellyfish, corals, and sea anemones to capture prey.

The lesson? I suppose that this story of miniaturization pushes the limits of the awesome things that living organisms can do. So much for thinking of unicellular organisms as being "primitive." And yet the question still haunts one: why did these multitalented protists not follow the multicellular path?

This subject was discussed on the podcast This Week in Microbiology #103: The battle for iron (bit.ly/1MrFhjM).

May 10, 2015

bit.ly/1OFyram

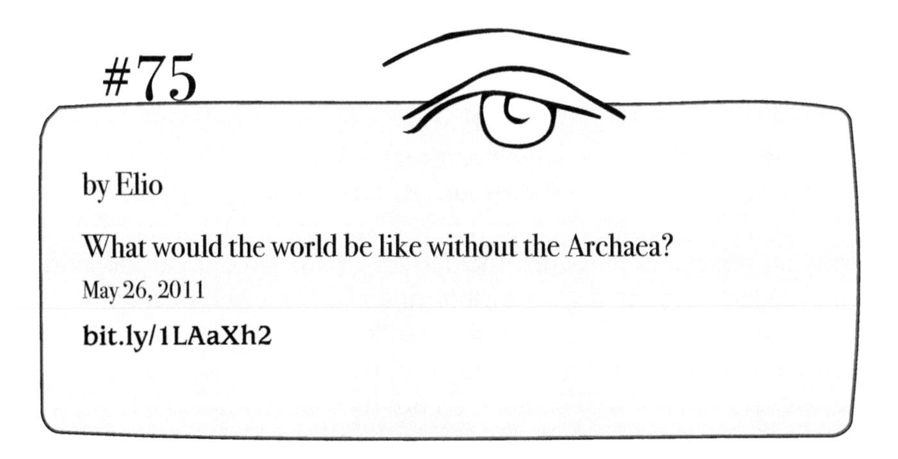

#75

by Elio

What would the world be like without the Archaea?

May 26, 2011

bit.ly/1LAaXh2

33

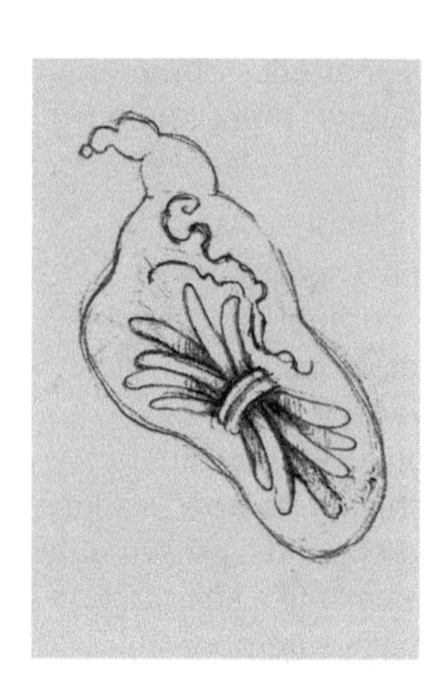

Let's Not Forget *Acetabularia*

by Elio

If I asked you what was the experimental basis for the central dogma of biology (DNA makes RNA makes protein), you would be likely to mention the classical findings that the transforming principle was DNA (Avery et al.) or that phages transfer DNA to the host (Hershey & Chase). However, it is unlikely that you even have heard that the precept was earlier derived from studies with a unicellular marine alga, *Acetabularia*. If so, you would miss the remarkable biology that made it possible to carry out this work. Here is why: *Acetabularia* is such a large cell that it can be readily handled with one's hands. It can be amputated into pieces that can be grafted together and its nucleus transplanted as easily as walking in the park.

Most cells are clearly too small for such luxuries. To enjoy them, we must turn to the outliers in range of sizes, that is, to giant cells. So, how big can cells get? The champion seems to be another a marine alga, *Caulerpa*, which can reach 3 meters in length. It is multinucleated, which seems almost like cheating (consider acellular slime molds, which can also reach enormous sizes, and other coenocytic organisms). Incidentally, Caulerpas are edible and are called sea grapes in Okinawa (umi-bud). Also multinucleated are the xenophyophores, foraminifera-like protists that live in the ocean at depths below 500 meters and reach 15 cm across. Among the largest uninucleated single cells are the foraminifera called Nummulites, which can reach 5 cm in diameter, and a marine amoeba called *Gromia spherica*. But the giant-celled algae, including

Acetabularia, are high on this list. Of course, birds' eggs are even larger and they are unicellular and uninucleated all right, but their contents are more suitable for an omelet than for experimental manipulations.

But first, what are Acetabularias? They are umbrella-like green algae (called, probably by some Victorian era naturalist, "*mermaid's wine glasses*") that are found in subtropical seas. They will remind you of the leaves of nasturtium, should you remember what these look like (entirely round leaves atop central stems). Acetabularias can reach an astounding 3 – 6 cm in height and consist of a slender stalk that is usually attached to a rock surface by a rhizoid and which ends in a lobate umbrella-like cap. A large nucleus (50 – 120 μm in diameter) is located at the rhizoid. This nucleus divides repeatedly as the alga matures, and the daughter nuclei are carried upward by cytoplasmic streaming to end up in each of the lobes of the cap.

A clump of Acetabularia acetabulum.

Credit: Albert Kok.

What made them famous? Acetabularias were used in the early days of cell biology for some amazing experiments on nuclear transplantation and cellular development. About 75 years ago, Joachim Hämmerling discovered that he could cut the organism in half with impunity and that each half would regrow the alga. He also found that if he cut it again, the specimen that had come from the top half would wither away and not regenerate further, whereas the one from the bottom half did regrow even after many such surgeries. Is the reason that the bottom part has the nucleus? He answered the question by a direct experiment, based on the ability of the cut upper pieces to be grafted onto lower pieces with the nucleus. The grafts took and a new plant regenerated, cap and all. The two species Hämmerling chose differed in their cap morphology, one, *A. crenulata*, having a wavy-edged or crenulated cap, the other, *A. mediterranea* (now called *A. acetabulum*), a smooth one. The defining observation was that the regenerated plant had the morphological characteristic of the bottom part, which was the first conclusive demonstration that the nucleus controls development. To confirm this, he transplanted nuclei of one species into another and got a hybrid with characteristics of both species. Hämmerling's experiment was the forerunner of all subsequent nuclear transplantations, which started some 20 years later with John Gurdon's experiment showing that a mature frog could be produced by transplanting a nucleus from an intestinal epithelial cell into an enucleated egg. For a video of Hämmerling's experiment, see here (bit.ly/1O6gmm1).

Hämmerling discovered another surprising fact: the anucleated top portions of the cut stalk could not only regenerate the caps (albeit, as mentioned already, only once), but those pieces farthest removed from the nucleus regenerated the best caps. This suggested that the stalk contained substances involved in cap morphogenesis that were distributed along a gradient, the lowest concentration being at the basal end near the nucleus and increasing toward the top of the stalk. He also found that such substances were involved in the formation of all the parts of the organism, including whorls of hairs towards the top of the stalk and the rhizoid. In later experiments involving interspecific grafts, Hämmerling and colleagues showed that the nucleus is the source of these morphogenetic substances.

How did Acetabularias unravel the central dogma? In his day, Hämmerling could not guess the morphogenetic substances were messenger RNA (which was not described until 1961), but he soon declared that they "*carried genetic information from the nucleus to the cytoplasm.*" He thus anticipated the discovery of mRNA by some 30 years. In time, he and several other investigators established that these substances were indeed mRNAs. Later on, Jean Brachet and colleagues in Brussels studied RNA and protein synthesis in nucleated and anucleated *Acetabularia* and showed that RNA flows from the nucleus to the cytoplasm, where proteins are synthesized and ultimately account for the differentiation of umbrellas. (In the 1950's, Brachet's lab was one of the European meccas for Americans to do a postdoc.)

What is the current status of *Acetabularia* research? Work with this model organism has been helped greatly by manipulations that shorten its life cycle. The life cycle used to take some six months in the lab and one to two years in the wild. A series of improvements in the medium and the use of axenic zygotes plus other tricks shortened the life cycle to about 97 days, a big jump, considering. This is not *E. coli*, but going from zygote to zygote in that time is not bad compared with "higher" plants and animals.

Although its aficionados may well argue that *Acetabularia* has not received its proper share of attention of late, considerable work has been carried out with it in recent times. As you would expect, its giant size opened the door to studying details of how its constituents are distributed along its various regions. For example, it has been shown that specific mRNAs are distributed along the stalk and that they fall into four classes of localization: throughout the organism, at the base, at the apex of the stalk, and in locations that change during development. These molecules are likely transported via actin microfilaments of the cytoskeleton that span the whole cell. *Acetabularia*'s cytoskeleton consists of parallel bundles of actin that run axially and are involved in forceful cytoplasmic streaming. Also well studied is the behavior of the various classes of mRNAs involved in the synthesis of structural and regulatory proteins. It is noteworthy that some, notably the mRNAs for tubulin, are made and stored long before being used. How is this done? Good question.

Much information has become available regarding cytoskeletal and endomembrane dynamics, electrophysiological elements affecting ion fluxes, and synthesis and mechanical properties of the cell wall. Like elsewhere, signal transduction and hormonal control are involved in regulatory pathways. In addition, the effects of environmental factors including light and gravity have been studied extensively and found to be involved in regulation of morphogenesis.

Open for more work with this organism are such fundamental questions of developmental biology as how structure and function operate at localized regions and how they are established and maintained. In the words of D. F. Mandoli, one of the principal present-day investigators of this organism: "*to me, the special appeal of A. acetabulum lies not just in the ability to address important questions in developmental and structural biology in the context of a physically large and architecturally complex unicell, but in being able to do so with access to a diverse and robust toolkit because this means that if one avenue of attack does not work, another probably will.*"

January 11, 2015

bit.ly/1W1YgQQ

by Elio

Why are there so many species of microbes on earth?

August 25, 2011

bit.ly/1LAaXNS

34

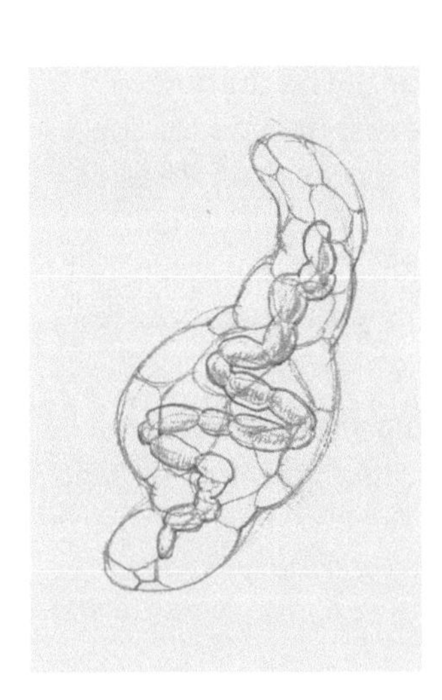

The Fastest Flights in Nature

by Elio

If you were a fungus growing on dung and if, for propagation, you depended on having your spores ingested by an animal, you'd have a problem. Few animals graze near their poop. The solution would be to evolve a powerful discharge mechanism that ensured that your spores traveled far afield. In the words of a pioneer in this area, Nik Money: "*Evolution has overcome this obstacle by producing an array of mechanisms of spore discharge whose elegance transforms a cow pie into a circus of microscopic catapults, trampolines, and squirt guns*."

A paper from Money's lab at Miami University in Oxford, Ohio, and colleagues at the College of Mount St. Joseph in Cincinnati, Ohio, reveals how fungal squirt guns fire spores over distances of more than 2 meters. The researchers used high-speed cameras running at up to 250,000 frames per second to capture these blisteringly fast movements.

The videos were made at camera frame rates of up to 250,000 fps. Each discharge is completed in the Anvil Chorus from Verdi's *Trovatore*.

Spores are launched at maximum speeds of 25 meters per second—remarkable for a microscopic cell. This corresponds to accelerations of 180,000 g, no less! In terms of acceleration, *these are the fastest flights in nature*. The fungi studied were two ascomycetes (*Ascobolus immersus* and *Podospora anserina*) and two zygomycetes (*Pilobolus kleinii* and *Basidiobolus ranarum*).

How do these fungi do it? Fungal cells make use of osmotic

pressure, but just how is the pressure generated? The authors identified several sugars and other osmolytes responsible for the water influx into the guns. The long flights of spores do not result from unusually high pressure but from the explosive way the pressure drop is harnessed for spore propulsion. There appear to be some similarities to the expulsion of ink droplets through nozzles on inkjet printers.

As the authors state in a press release: "*This information is very important for future biophysical studies on spore and pollen movement, which have implications for the fields of plant disease control, terrestrial ecology, indoor air quality, atmospheric sciences, veterinary medicine, and biomimetics.*"

We thank Nik for providing us thc movic. Wc had featured his work on the discharge of spores in basidiomycetes (which include mushrooms) previously. An interview with Nik broadcast on CBC radio is also available.

Sporangiophore discharge in *Pilobolus kleinii* captured with high speed video. This illustration is a montage of 6 images from a video obtained at a frame rate of 50,000 frames per second. The selection shows every 10th frame, representing 200 microsecond intervals. The launch is completed in < 0.25 milliseconds or < 1/400 of an eye blink.

Source: **Yafetto L, Carroll L, Cui Y, Davis DJ, Fischer M, Henterly AC, Kessler JD, Kilroy HA, Shidler JB, Stolze-Rybczynski JL.** 2008. The fastest flights in nature: high-speed spore discharge mechanisms among fungi. *PLoS ONE* **3**:e3237.

September 25, 2008

bit.ly/1QNQyIW

35

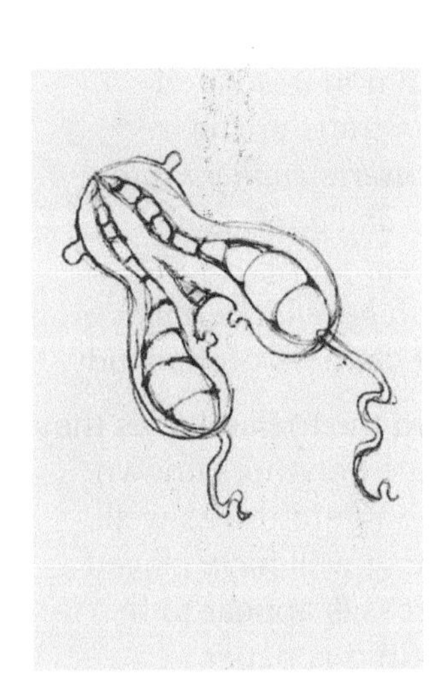

Fungal Alchemy: Using Radiation as a Source of Energy

by Elio

Accustomed as we are to microbial surprises, we were nonetheless taken aback by a report disclosing that certain fungi grow better when exposed to ionizing radiation. According to a paper from Albert Einstein Medical School, fungi can also use radiation as a source of energy—not exactly one's view of radiation as something malevolent and baneful. Had we paid attention to the news from Chernobyl, we would not have been surprised because the walls of the still-hot reactor have become covered with mold. Not only that, but it has been known for some time that fungal species search out radioactive particles, i.e., manifest radiotropism.

Many fungi are relatively radioresistant, apparently due to their melanin content. These species are easy to spot because they turn mildewy walls black and form black colonies on agar plates. (These dark fungi even have a term of their own: dematiaceous.) There is, of course, a big difference between being resistant to radiation and actually using it as a source of energy as is proposed in this study. Significant growth enhancement by radiation was seen in *Cryptococcus neoformans* (which becomes melanized when grown on suitable precursors) and naturally melanized species such as *Wangiella dermatitidis* and *Cladosporium sphaerospermum*, but not in albino mutants. These results suggest that melanin is the antenna that captures radiation and converts it into usable biological energy.

The authors present physicochemical studies showing that melanin is indeed able to capture radiation. As shown by electron spin resonance, radiation alters melanin's electron structure. Irradiated melanin reduces NADH four times faster than non-irradiated melanin. Thus, radiation increases its electron-transferring properties. "*Just as the pigment chlorophyll converts sunlight into chemical energy that allows green plants to live and grow, our research suggests that melanin can use a different portion of the electromagnetic spectrum ionizing radiation—to benefit the fungi containing it*," says Dr. Dadachova, one of the authors of the study.

There is a lot to melanin (or, more properly, to the melanins, as they come in quite a few varieties). They are unique polymers known to protect living organisms against extreme temperature as well as both UV and ionizing radiation. Melanin may well have played an important role in evolution. Many fungal fossils appear to be melanized—a point discussed in some detail in this paper.

A headline in the press release from the Albert Einstein College of Medicine modestly announces that their findings "could trigger recalculation of Earth's energy balance and help feed astronauts."

June 18, 2007
bit.ly/1W304yy

36

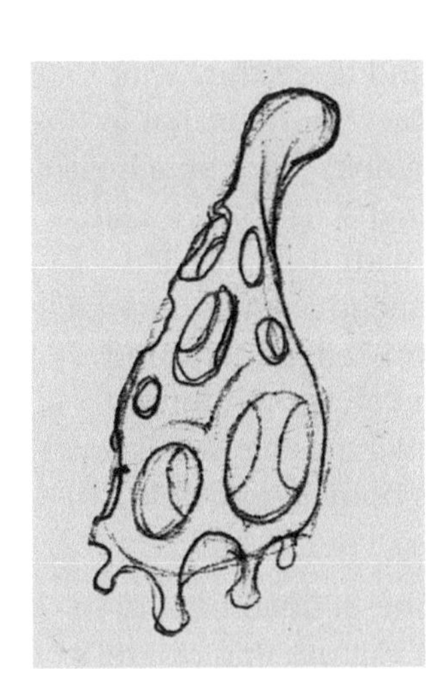

Arms and the Fungus

by Elio

"*Arma virumque cano*" ("*I sing of the arms and the man*"). So starts Virgil's *Aeneid*. I have wondered why anyone would want to sing the praise of armaments, as this is not my cup of tea. There is, however, biological weaponry that I think deserves our admiration. One of the most elaborate examples is the mechanism found in *Haptoglossa,* an enigmatic fungus that has been provisionally classified with the oomycetes or water molds. An entire fungal cell is injected into a passing nematode or rotifer, there to develop at the expense of the host. Mechanisms for injecting biologically active material are not uncommon; they are seen in pathogenic bacteria, viruses, jellyfish, and stinging nettles, among others. But it is a rare organism that possesses the intricate machinery required to actively introduce a whole cell into a fast-moving and unsuspecting host.

An apology first: I use anthropocentric talk here, not because it's the right thing to do, but because it is convenient. Unicellular *Haptoglossa* motile spores (zoospores) differentiate in a few hours into a fancy structure called a "gun cell," which is capable of injecting a projectile—a spore—*in toto* into a grazing rotifer or a nematode. Once inside, the spore develops into a vegetative fungal body that consumes the host and grows to fully occupy its body cavity in a few days. Motile spores are then produced and released from the carcass of the host, starting the cycle anew. Thus, *Haptoglossa* is an obligate parasite and infection is required for survival.

What are the steps in the firing of a gun cell? Gun cells typically anchor to the substrate by their sticky base, and there they wait for an unsuspecting small animal to happen by. When a suitable host touches the business end of the gun, it is held there by a host-specific adhesive. *Haptoglossa,* by the way, means "sticky tongue." Injecting the projectile requires close contact with the host. This is not like a bullet flying through space. After attaching, the prepared projectile is shot through the muzzle of the gun cell, penetrating the animal's epidermis. A tube inside the gun cell is then turned inside out, forming a narrow conduit through which the *Haptoglossa* nucleus and cytoplasm pass into the animal's body. The whole process takes but a fraction of a second.

The structural differentiation of gun cells has been documented in great detail. The image of a section through a mature gun cell tells part of the story. One sees what looks like a wicked harpoon with a long shaft and barbs at its pointed end. Mournfully, it reminds one of the harpoons used to capture and kill whales by whaling ships. Scary-looking though this may be, the picture is slightly deceiving because the "barbs" are not hooks used to grab onto the host, but rather they are coiled proteins in a jack-in-the-box like arrangement. A similar arrangement in horseshoe crab sperm was recently shown to function as a biological spring capable of storing mechanical

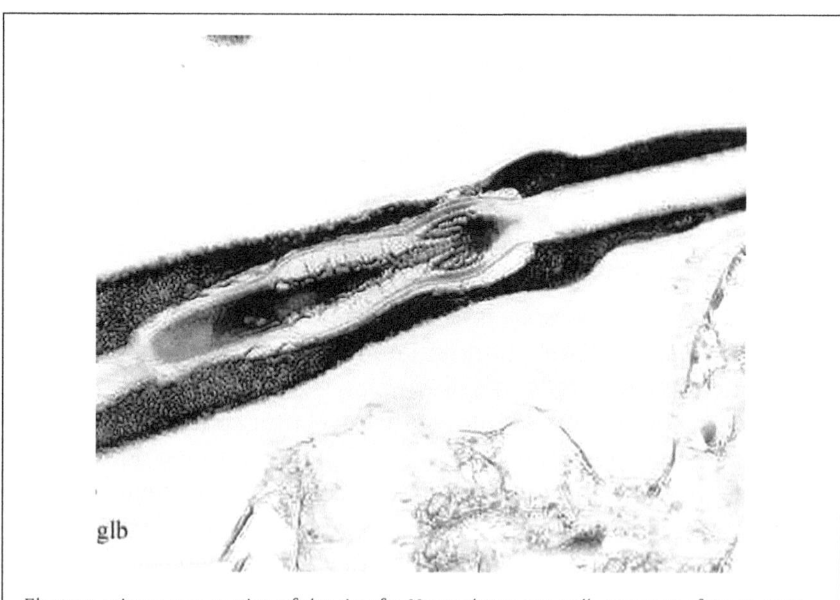

Electron microscope section of the tip of a Haptoglossa gun cell, courtesy of Dr. E. Jane Robb.

energy. Release of this energy to do work requires neither the action of motor proteins nor actin polymerization. It seems possible that *Haptoglossa* may use such a spring mechanism to fire its projectile. On the other hand, some researchers suggest that firing results from the rapid build up of osmotic pressure in the swollen basal part of the gun cell, and still others postulate contraction of actin filaments triggered by a rapid influx of calcium ions. The debate and the research continue.

Of course, *Haptoglossa* is not alone in its ability to penetrate into other living cells. Sperm do it. Protozoa do it. Even bacteria and viruses do it. Each has its own mechanism to effect penetration, some simple-sounding, others quite intricate. Most require at least some active participation by the host. *Haptoglossa* (and some of its distant relatives) beats them all for the intricacy and beauty of its engineering design. I, for one, stand in awe of such structural sophistication. Someone should set *Haptoglossa*'s performance to music.

March 21, 2007

bit.ly/1OFzh78

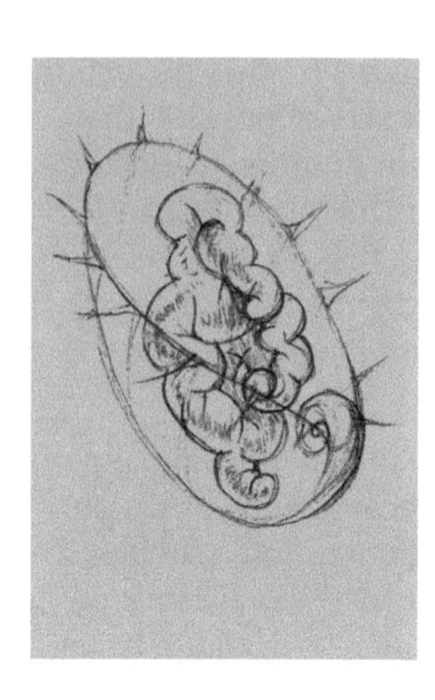

37

Bacterial Hopanoids: The Lipids That Last Forever

by Elio

The world of lipids does not always gets its due. Their oleaginous charm is not always appreciated. For example, have you heard of hopanoids? They are made by some bacteria and are an unusual kind of polycyclic lipids that resemble steroids, but with an extra ring. Just like cholesterol in eukaryotic membranes, they insert in bacterial membranes where they contribute to their stability. Both hopanoid and steroid molecules are almost planar, thus can intercalate into the lipid bilayers with relative ease and there interact with the adjacent fatty acid chains to enhance stiffness. That hopanoids and steroids play analogous roles can be readily shown in mycoplasma, the bacteria that have the unusual ability to incorporate exogenous steroids in their membranes. *Mycoplasma mycoides*, it turns out, can grow without steroids if provided with hopanoids. Not all bacteria make hopanoids, but they play a vital role in the ones that do. How do we know? Inhibitors that stop their synthesis also inhibit the growth of the organisms that make them. Hopanoids are found in trace amounts in some plants and not at all in the Archaea. For a review, see here (1.usa.gov/1H96TT8), and for previous appearances in this blog, here for a piece (bit.ly/1MrGNCg) by Tanja Bosak and here (bit.ly/1MrH4VO) for one by Paula Welander.

Hopanoids are complex molecules, decked out with a large variety of side chains characteristic of the species that makes them. Why such a diversity? Not enough is known about this to be able to figure out if there are rules that govern this. Is this related to requirements imposed by their environment? Puzzling is that within closely related species, some make them and others don't. For example, some nitrogen-fixing bacteria in the genus *Azotobacter* produce hopanoids but close relatives (e.g., *A. chroococcum*) don't.

More is known about the hopanoids physiological roles. They help membranes withstand damaging stress conditions, among others high temperatures, low pH, and detergents. Hopanoids usually comprise around 1–5% of the cell's total lipids, but in some cases their proportion raises considerably in response to stress. Thus, they make up about 50% of the lipids in *Zymomonas*, a bacterium that vies with yeast for a high level of alcohol production. More examples: the thermophile *Bacillus acidocaldarius* makes about seven times more hopanoids at 65° C than at 60° C. At high temperatures, hopanoids may counteract the increased fluidity of the lipid portion of the membrane and thereby reinforce it. One interesting use for hopanoids is seen in the actinomycetes, bacteria that make filaments that stick out into the air. This aerial mode of life calls for special membrane properties that may be related to their hopanoid content. Moreover, hopanoids have been seen in *Bacillus subtilis* in the act of sporulation.

There is something well nigh incredible about hopanoids and steroids: they are found in very ancient rocks. The hopanoid rings are amazingly stable to acid and alkali conditions and high temperatures, so that when rocks and petroleum were formed, they were among the few organic molecules that survived. This may have taken place as long as 1.64 billion years ago. In the process, hopanoids were stripped of some of their side chains, leaving a simple hydrocarbon skeleton behind. With all this time in the history of the planet, hopanoids accumulated in spectacular amounts, perhaps as much as 10^{12} tons. This makes them enormously abundant and equal in mass to the organic compounds of all organisms now living. Almost certainly, they represent the largest mass of any single class of organic molecules on Earth. And yet, they are unfamiliar to most people, in and out of science.

Hopanoids—especially with a 2-methyl substitution—serve as biomarkers, that is, as indicators of the extinct organisms that existed at a given geologic time. Notice how useful it could be to be able to rely on such markers, especially because the evidence based on the morphology of fossils is often controversial. The notion has been around for some time that 2-methyl hopanoids tell us that the earliest photosynthesizing microbes, the cyanobacteria, were present at the time rocks were formed. Given the importance attributed to ancient cyanobacteria in forming atmospheric oxygen, this is a key event in the history of the Earth.

But there are problems with this idea. Some bacteria other than the cyanobacteria also make hopanoids while growing anaerobically, thus not producing oxygen. In addition, cyanobacteria differentiate into resting cell, "akinetes," and these produce ten times the amount of hopanoids as vegetative cells. They are found in greatest amounts in the akinete outer membrane (which is also the preferred localization in other bacteria). This suggests that hopanoid protection kicks in when cells become desiccated or exposed to prolonged cold, which does not support an exclusive role for them in photosynthesis. Thus, they do not serve too well as biomarkers for ancient oxygenic photosynthesis. All is not lost: they may serve as indicators of other biological attributes. Their presence may indicate that olden-day cyanobacteria were already capable of differentiation into akinetes, thus pointing to this being an ancient skill indeed.

Looking for something else that hopanoids may tell us about early life does not stop there. The group of Ricci and colleagues (1.usa.gov/1WhEscv) went about it in a systematic way and assayed for 2-methylhopanoids in a large number of bacteria. Instead of looking for hopanoids directly, they assayed metagenomes from various terrestrial and aquatic sources for *hpnP*, the gene encoding C-2 hopanoid methylase, and *shc*, the gene for squalene hopane cyclase, both involved in their synthesis. Indeed, these genes were found mainly among the alphaproteobacteria. Surprisingly, most cyanobacteria, once thought to be great 2-methylhopane producers, do not have the *hpnP* gene. Perhaps the ancient cyanos did, but the data do not encourage such a view.

So, how common are these hopanoid-related genes? They are relatively infrequent, being seen in only about 4% of the metagenomes analyzed. The majority, 63% of all these, are represented by terrestrial species. This conclusion should not make paleomicrobiologists particularly happy because although the fossil record suggests that 2-methylhopane producers lived in shallow tropical seas, they are not enriched in such habitats today. Today, genes for 2-Me-hopane is found in a great many habitats, which does not help in the interpretation. But one fact emerges: about half of the hopanoid formers are associated with plants in a mutualistic or commensal association. Note that plants evolved some 500 million years ago, so this has little to do with the oldest findings of hopanoids. But these habitats, ancient and not so ancient, may have something in common: they supported sessile bacterial communities (biofilms?) that are at low oxygen concentrations. The paper discusses various types of biases in such studies, concluding that the main interpretations still hold.

In conclusion, hopanoids and their role in evolution need more work, which is easy enough to say. Like in all of science, as older ideas wane, new ones take their place. In this spirit, the authors say: "*Our ecological data demonstrate that 2-MeBHPs (bacteriohopanepolyol) cannot be used as taxonomic biomarkers for any particular group but suggest 2-MeBHPs may be diagnostic for the confluence of particular environmental parameters.*"

In the words of Tanja Bosak in these pages (bit.ly/1MrGNCg): "*In the beginning, there were fats, and in the end, only fats will remain.*"

References

Ricci JN, Coleman ML, Welander PV, Sessions AL, Summons RE, Spear JR, Newman DK. 2014. Diverse capacity for 2-methylhopanoid production correlates with a specific ecological niche. *ISME Journal,* **83**:675-84.

September 28, 2014

bit.ly/1jy7kkI

38

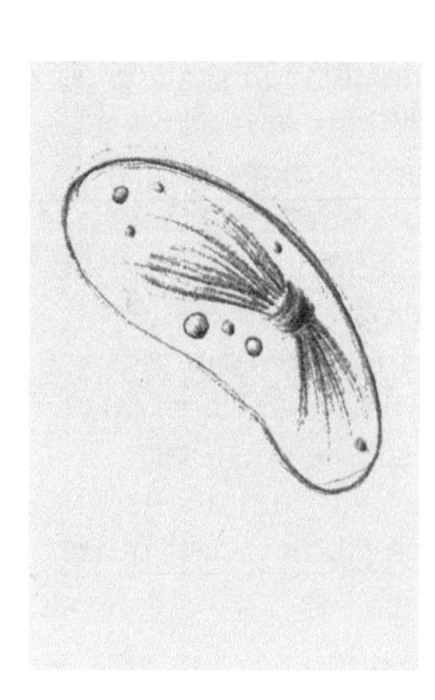

A Bacterium Learns Long Division

by Nanne Nanninga

The common picture of a dividing rod-shaped bacterium encompasses the positioning of the divisome, including an FtsZ-ring, in the cell center. This occurs after the cell has doubled its length without increasing its diameter. Conversely, increase in diameter without cell elongation would seem highly unlikely in a rod-shaped organism. Yet, this happens.

In fact, this is the normal condition for an ectosymbiotic

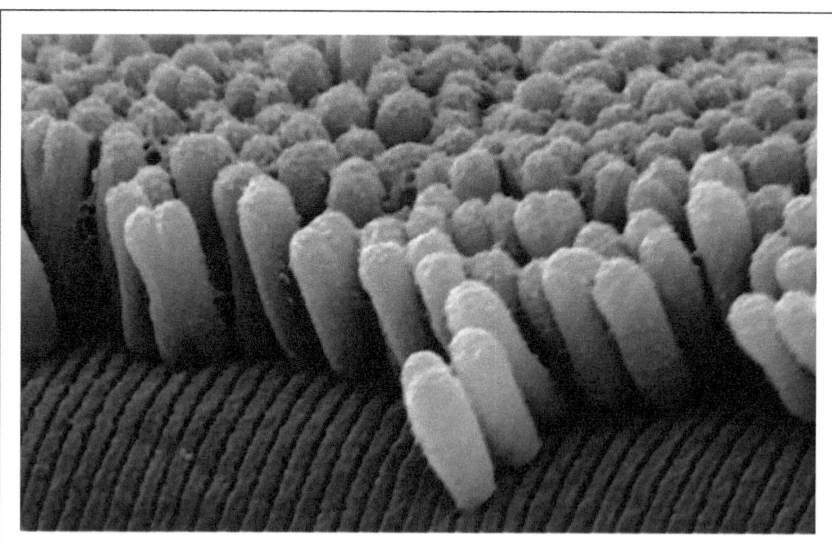

Fig 1. SEM of tightly apposed ectosymbionts on the surface of *L. oneistus*. Electron micrograph by N. Leisch.

bacterium that lines the surface of the marine nematode *Laxus oneistus*. The symbiont is a gammaproteobacterium like *E. coli*, but unlike *E. coli* it has not been cultured outside its natural habitat. As originally described by Polz et al. (bit.ly/1ifjjCj) in the early nineties, the bacteria are positioned perpendicular to the surface of *L. oneistus*. They are glued to the surface of the nematode by a C-type lectin. The original paper already indicated that division takes place longitudinally, with one of the daughters presumed to remain attached to the nematode. This makes sense because otherwise the daughter cells would get lost to the environment. Recall that this phenomenon lies at the basis of the Helmstetter-Cooper baby machine, whereby one of a pair of newborn cells is selected to start a synchronous culture.

Recently, Leisch and colleagues (1.usa.gov/1N6t5DI) have extended these observations by carefully determining cellular

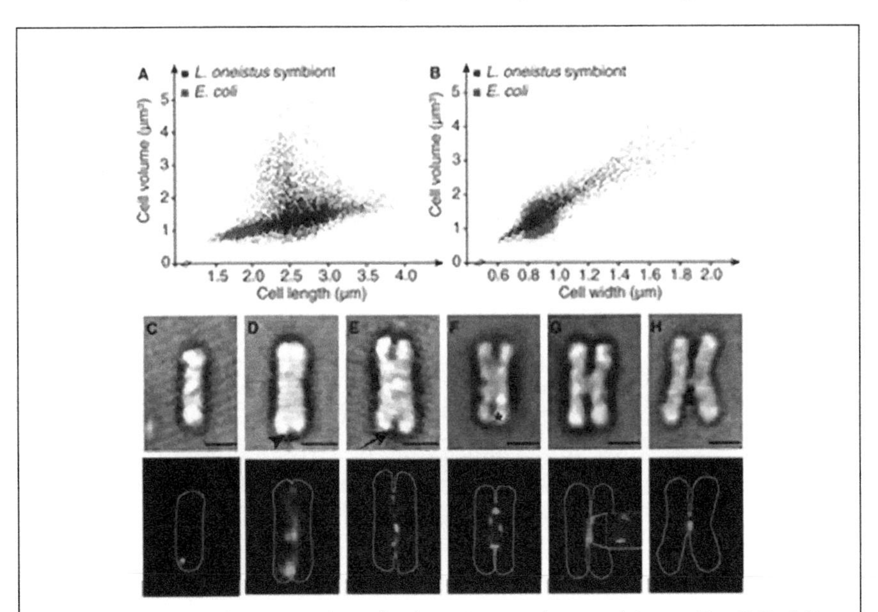

Figure 2: A. Cell length compared to cell volume in *L. oneistus* symbiont or *E. coli*. **B**. Cell width in *L. oneistus* or *E. coli* compared to cell volume. **C-D**. Cells arranged according to division stage and imaged by differential contrast images (upper row) and confocal fluorescent microscopy (lower row). FtsZ is visible as green spots in the lower panel.

Source: http://www.ncbi.nlm.nih.gov/pubmed/23058799

dimensions and visualizing the FtsZ division protein with fluorescent *E. coli* monoclonal antibodies. The results can be compared with *E. coli* data (Figure 2A, B). Whereas *E. coli* elongates as expected, this is not what happens with the symbiont. The symbiont does not change its length (Figure 2A) but increases its diameter (Figure 2B). Consequently, the symbiont divides longitudinally (Fig. 2 C-H). Immunostaining of FtsZ reveals that FtsZ positioning correlates with the cell constriction. In fact, an ellipsoidal FtsZ ring is observed stretched along the length of the endosymbiont.

These remarkable observations provide more questions than answers. For instance, how does the diameter of the symbiont increase? Is the increase the same in all directions, or is it polarized? Though, the authors do not discuss this point, it would seem that the bacterium has the shape of shoebox. Thus, increase in width implies widening of the shoebox without altering its height. How is the shoebox-shape, if applicable, maintained? Furthermore, how is DNA segregation carried out in the absence of cell elongation? How are the nucleoids arranged? It would seem that widening of the shoebox is sufficient. Again the spatial mechanism is not known. As pointed out by the authors, genes encoding MinC, MinD, and MinE are also present. Do they function in the symbiont, and if so, how do they conform to different geometry? Answers to these questions are likely to elucidate the degrees of freedom in bacterial cell division. But that's what we learned in our arithmetic class about the long division. Unless one deals with a whole number, there is always a remainder.

Nanne Nanninga is Emeritus Professor of Molecular Cytology, Swammerdam Institute for Life Sciences, University of Amsterdam, The Netherlands.

November 15, 2012

bit.ly/1RIla4k

39

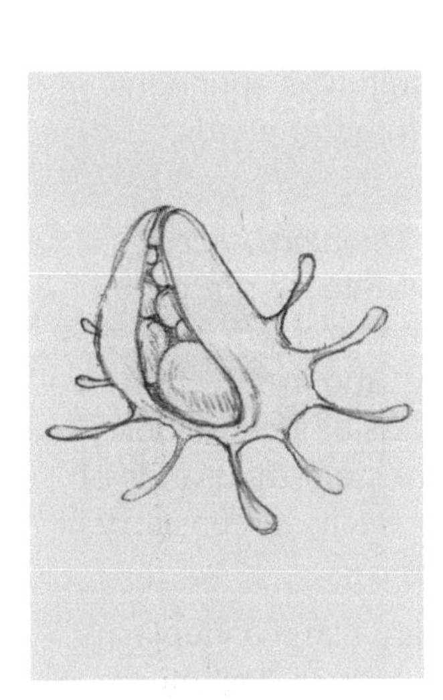

Commuting to Work

by Elio

An underwater microbial mat has been found in fairly shallow waters off the coast of Chile and, according to headlines, it's the size of Greece, or about 132,000 km^2 (or for us norteamericanos, about the size of Alabama). These communities of sulfide-oxidizing bacteria have been known for some time, but their attention has been highlighted by the recent version of the Census of Marine Life. In fact, they were discovered in 1963 by the oceanographer and microbiologist Victor Gallardo of Chile's University of Concepción. Scientists may not have known much about these huge mats much earlier on, but the local fishermen sure did and called them *estopa*, Spanish for burlap or unwashed wool or flax. Aficionados of the giant sulfide-oxidizing bacterium *Thiomargarita namibiensis* and other bacterial gargantuas likely include *Thioploca*, the occupants of these mats, in their catalog of microbial marvels. This is a genus of gliding, filamentous bacteria that live in aquatic sediments where they face the same problem as *Thiomargarita*, namely, how to hook up their fuel (sulfides) with their final electron acceptors (nitrates). (For details of their metabolism, see a recent paper [1.usa.gov/1RAhxrq].)

Sulfides are abundant in the sediment, nitrates in the water column. How to bring the two together? Thioplocas solve the problem by making tubular sheaths that stick out from the ocean's sediment. These sheaths can be as long as 15 cm. Inside them, filaments of the organisms glide up and down, gathering sulfides below and nitrate above. Thioplocas chemotax towards nitrate. They absorb it and transport it downwards, to where the sulfide is abundant, at speeds of about one centimeter per hour. In other words, Thioplocas take an elevator to work.

The two dominant species in these mats are *Thioploca chileae* (with cells 12 to 22 μm wide) and *T. araucae* (28 to 42 μm wide). Their cells, 60 μm long, are arranged end-to-end to form filaments (*trichomes* in the jargon) some 2 to 7 cm long, and arranged in parallel bundles within the sheath. Each cell has a huge vacuole filled with 0.5 M nitrate (that's about a 4 % solution, thousands of times the concentration in sea water) that serves as a repository of oxidant, pretty much like a scuba diver's tank.

The giant size of the organisms does not mean a large cytoplasm; most of the cell is a vacuole, just as in *T. namibiensis*. Thioplocas are close relatives of the *Beggiatoa*, another group of giant bacteria that oxidize sulfides and also make giant mats in marine and freshwater environments.

The Thioploca's gelatinous sheaths get to be huge, walls as thick as 1.5 mm and as long as 10-15 cm, thus readily seen with the naked eye. The bottom 5-10 cm is buried in the sediment, thereby

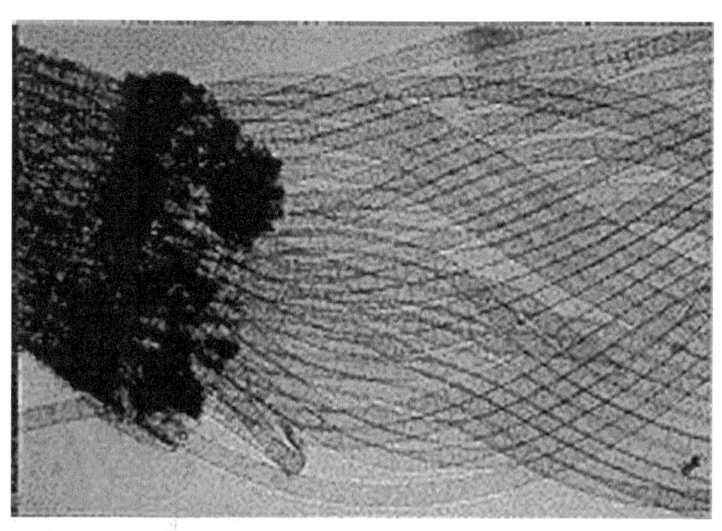

Bundle of *T. araucae* extending out of their sheath. The appearance of a braid is seen where the filaments cross over, hence the name *Thioploca*, "sulfur braid."

Source: **Jørgensen BB, Gallardo VA.** 1999. Thioploca spp.: filamentous sulfur bacteria with nitrate vacuoles. *FEMS Microbiol Ecol* **28**:301-313.

firmly anchoring the sheaths and making the mats very stable. The sheaths lie flat along the surface and look yellowish, according to those who have seen them. Each sheath may be shared by as many as 100 individual filaments, each of which can glide up and down independently.

I have tried to find out more about the nature of the sheaths, but have had little success so far (please inform me of what I may have missed). It is said that they consist of mucilaginous polysaccharides. But think about it. How can a bundle of bacteria make such a thick extracellular tube of the proper dimensions? Which comes first, the bundles of cells or the sheaths? The sheaths appear to be made up of layers; how do these layers get deposited? The outside of the sheaths seems to pick up minute pieces of detritus. Is the surface sticky? And is there someone in the vicinity who likes to munch on these probably nutritious sheaths? More questions, please.

It gets even more interesting. Other inhabitants of the mats also live in the sheaths, especially the upper portions. These include filamentous sulfate-reducing bacteria, smaller than the *Thioploca*, thought to be members of the heterotrophic genus *Desulfonema*.

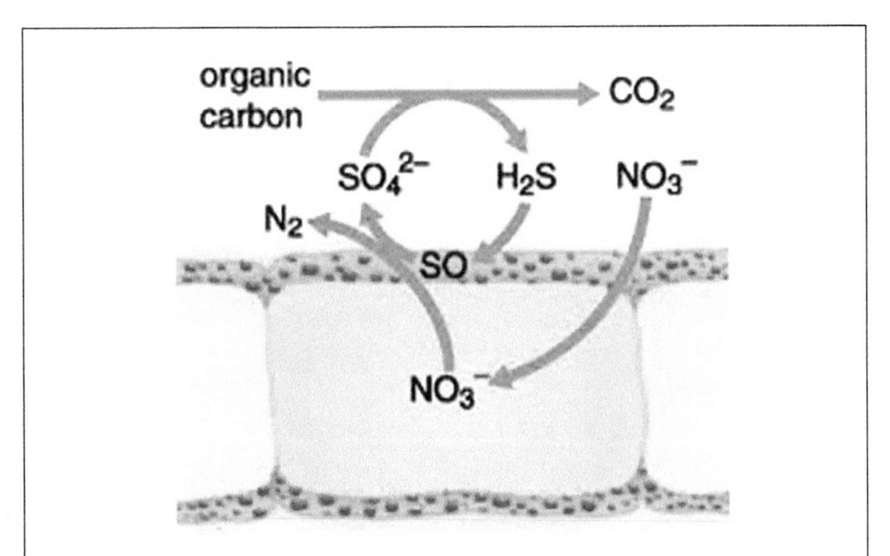

A schematic drawing of a *Thioploca* cell showing the large vacuole that occupies nearly the whole space and that contains nitrate at concentrations up to 500 mM, some 20,000 times that of the surrounding seawater.

Source: **Levin, LA.** 2002. Deep-ocean life where oxygen is scarce. *American Scientist* **90**:436-444.

At other oceanic sites, yet other "endosymbionts" are found with the Thioploca where, it is thought, they contribute to sulfur cycling within the sheaths by providing at least some of the sulfide needed by their larger hosts.

The total biomass of these mats is staggering. A square meter of mat at its densest weighs nearly one kilogram (wet weight), some 90 % of which is due to the sheaths. The extent of these mats is not known, although they are thought to occupy thousands of square kilometers, possibly spanning an area from Chile to Columbia—the length of South America. Note that the weight of the mat in a single square kilometer may approach 1000 tons! Or, if you prefer African elephant-equivalents, the weight of about 200 such pachyderms/km^2. Whatever metric you prefer, this is one of the largest biomass entities in the oceans and certainly one of the biggest accumulations of microbial life in any one place. Think of it next time you visit Greece.

References

Teske A, Jørgensen BB, Gallardo VA. 2009. Filamentous bacteria inhabiting the sheaths of marine Thioploca spp. on the Chilean continental shelf. *FEMS Microbiol Ecol* **68:**164–172.

Høgslund S, Revsbech NP, Kuenen JG, Jørgensen BB, Gallardo VA, van de Vossenberg J, Nielsen JL, Holmkvist L, Arning ET, Nielsen LP. 2009. Physiology and behaviour of marine Thioploca. *ISME J* **3:**647–657.

September 27, 2010

bit.ly/1LAqmxZ

PART 4

On Being a Microbiologist

Ah, the human
side of science!
We can look at it
in so many ways…
I present a few.

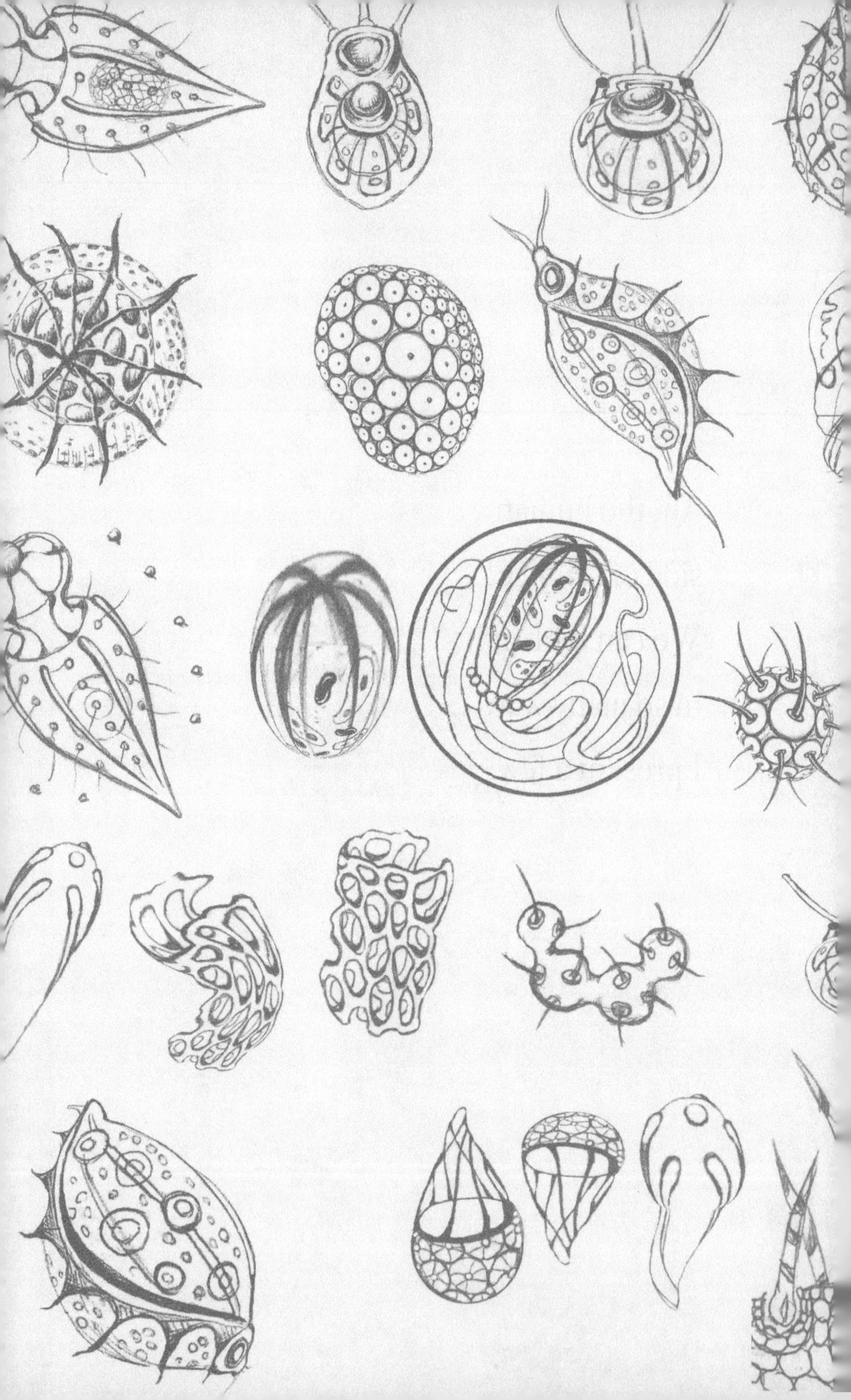

40

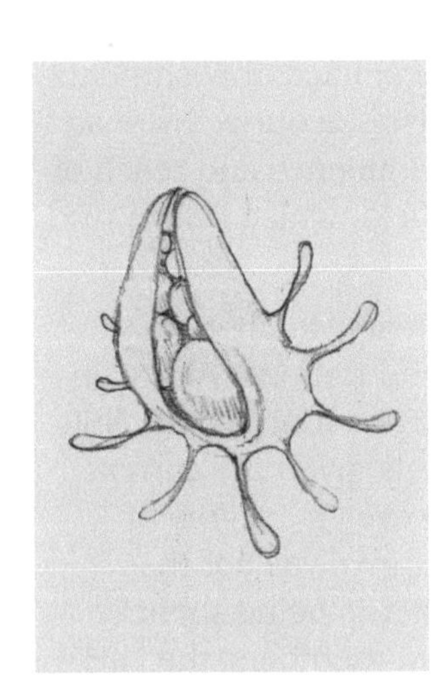

Directed Science, Curiosity-Driven Science, and Striking the Balance

by Jeff F. Miller

After nearly three months as president and one year as president-elect, what amazes me most about the ASM is the breadth and depth of our activities. We literally span the globe, with ambassadors in 54 countries and outreach programs in Africa, South America, and Asia focused on building laboratory capacity for tuberculosis, HIV, and other infectious diseases. The 14 journals we publish are reinventing themselves to remain vital in an electronic world, and *mBio*, our newly launched premier publication, scored an impressive impact factor of 5.3 less than two years after publishing its first issue. Our membership numbers are up nearly 2% from last year, with the largest increases in student and international members, and a new tiered structure will make it easier and more attractive to join. The General Meeting has recently undergone a metamorphosis, and I am pleased to report that total registration for 2012 was the highest since 2008, with over 9,300 attendees. The Academy of Microbiology continues to examine ways that microbes can positively impact environmental and human health, as exemplified by their recent colloquium on *Designing Drugs That Last*, and our Public and Scientific Affairs Board has been working tirelessly to advocate on our behalf to agencies and lawmakers throughout Washington. Their current efforts to lobby against across-the-board cuts in discretionary spending, as stipulated in the Budget Control Act of 2011 and slated to be enacted on January 2, 2013, are critical for preventing a potentially crippling blow to the nation's scientific research enterprise. This is just a small sampling of recent accomplishments and activities, and for more information please see our website at www.ASM.org. It is intended to illustrate the influence and reach of the ASM and its potential for advancing our field, which brings me to the point of this editorial.

As a basic scientist, there are few issues of greater interest or relevance than federal support for research and training. Although we would all embrace major increases in budgets for the NIH, NSF, and other funding agencies, this seems unlikely given the current climate, and maybe a bit naive. There is, however, a "degree of freedom" that deserves to be explored, and it involves not the absolute number of dollars, but how the money is being spent. To pick an example of relevance to many ASM members, the NIH recently released a concept statement for a program designed to build on the Regional Centers of Excellence for Biodefense and Emerging Infectious Diseases Program, which will sunset in 2014. The initiative will support multi-investigator, multidisciplinary translational research centers focused on generating, validating, and advancing therapeutics and vaccines, with significant oversight by program officers. This is admittedly an extreme case, but it raises two general issues. The first involves "managed big science," for want of a better term, and the second involves the balance between basic research and applications.

The need for transdisciplinary, multi-investigator programs that tackle big questions in biology is clear and well justified. The breathtaking insights from the Human Microbiome Project provide a beautiful illustration of the central importance of microbes and microbiology, and they would not have been possible without a massively integrated effort. Big science is important, but it comes at a significant cost, and it seems reasonable to ask how these efforts should be designed, managed, and balanced to produce the greatest value from our research investment. Requests for proposals involving large teams of investigators and correspondingly large budgets have become commonplace. These are designed by federal agencies with specific goals in mind, and they are often highly prescriptive, precisely outlining topics and approaches to be included, and excluded, along with additional requirements and stipulations. Is this the best and most cost-effective way to facilitate collaborative science? Has the approach become too "top-down" and should opportunities be designed to let even better ideas for big science "percolate up" from the research community? What is the optimal balance between targeted research and investigator-initiated inquiry? And perhaps most importantly, are directed approaches the best way to marshal the creativity and innovation required to achieve grand challenges, or do they stifle it?

Returning to the biodefense example cited above, the proposed initiative will focus almost exclusively on applications in the form of therapeutics and vaccines. The importance of applications that benefit society are undeniable, and they are an essential component of the justification for publicly supported basic research. But there is a growing perception in the research community that the pendulum has swung so far in this direction that it places discovery at risk. Curiosity-driven research has provided and will continue to provide the foundation for future applications, and by extension our ability to innovate and compete as a nation. This issue was recently considered in an editorial (bit.ly/1P3FtV8) by Francis Collins, director of the NIH. His contention that basic research has not been compromised at the expense of translational efforts may engender debate, but his conclusion should not. In it he states: "*In this time of severe budget constraints, Americans need to know that today's basic research is the engine that powers tomorrow's therapeutic discoveries... and they need to hear it from all aboard the biomedical research ship.*" Herein lies the challenge. As any basic scientist that has tried to explain his or her work to a non-scientific audience knows, it is much easier to articulate the importance of a potential application than to convey the essential nature of pure, unadulterated, curiosity-driven research. But if we don't do it, who will?

Questions regarding the optimal balance between directed science and investigator-initiated inquiry, and applied vs. basic research, are undeniably complex. They impact not only microbiology, but life and biomedical sciences and beyond. As the largest single-life sciences organization in the country, and one that has earned tremendous influence and respect, I submit that the ASM is poised to contribute to this debate in an informed and productive way. In addition to continuing our vital efforts focused on advocating for support for NIH, NSF, CDC, and other agencies, I believe it is time to devote increased attention to understanding how available research dollars are being spent, and to considering how investments can be leveraged to insure that a robust pipeline of basic discoveries will be available to support applications of the future. If done right, our colleagues in Washington will likely pay attention, and they may even follow our advice. It is equally essential to use our communications capabilities to more effectively and broadly convey the importance and excitement of curiosity-driven research and how it facilitates translation. Our educational efforts can also be focused on providing students and scientists with the skills to explain what we do and why we do it to non-scientists, in a way that engenders understanding and enthusiasm as opposed to confusion. The ASM has remarkable resources to draw upon, and these efforts will benefit from the concerted expertise of our Public and Scientific Affairs Board, Communications and Press, the Academy of Microbiology, our Education and Publications Boards, and other arms of the society. ASM should also explore strategic partnerships with like-minded organizations that have similar goals and synergistic capabilities. In addition to asking for increased research funding, it seems prudent to focus on maximizing the yield from support we already have. The vitality of our field may depend on it.

Jeff F. Miller is Professor and Chair of the Department of Microbiology, Immunology and Molecular Genetics at UCLA, and past president of the ASM.

September 17, 2012

bit.ly/1OPtRFc

41

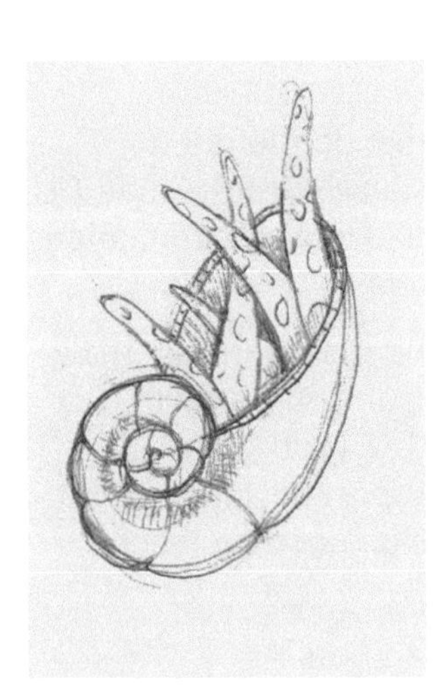

The Gender Bias of Science Faculty

by Vincent Racaniello

If you were a science professor, and you received two equally strong applications for the position of laboratory manager, one from a female, one from a male, which one would you pick? The answer (1.usa.gov/1WhFwwZ) may surprise you.

It is well known that women are underrepresented in many fields of science. Whether or not this disparity is a result of gender bias by science faculty has not been investigated. To answer this question, a randomized, double-blind study was conducted in which science faculty from research universities were asked to rate the application of a male or female student for a laboratory manager position. Identical applications were sent to all participants in the study, except that half ($n = 63$) received materials from a male student, John, and the others ($n = 64$) received materials from a female student, Jennifer. The faculty were then asked to rate the student's competence and hireability, and the amount of salary and mentoring that they would offer.

The results clearly show that the faculty felt that the female applicant was less competent than the male student, and offered Jennifer less career mentoring and less starting salary than John. Faculty gender, scientific field, age, and tenure status did not affect this bias. The data indicate that the female applicant was less likely to be hired because she was considered less competent than the male applicant.

What might be the reason for the subtle gender bias observed in this study? The authors suggest that it is due to a belief that women are less competent in science than men.

The fact that faculty members' bias was independent of their gender, scientific discipline, age, and tenure status suggests that it is likely unintentional, generated from widespread cultural stereotypes rather than a conscious intention to harm women.

I found the difference in mentoring offered the male versus female applicants most disturbing. An understanding and supportive mentor is an important component required for a successful career in science. Lack of encouragement and positive judgments may cause women to leave academic science before they reach university positions.

How can this subtle bias be eliminated? The authors suggest educating faculty and students about the existence and impact of bias within academia, an approach that has reduced racial bias among students.

We need more scientists in the US—one million over the next ten years, according to a 2012 report from the President's Council of Advisors on Science and Technology (1.usa.gov/1WgOiR4). Achieving this important goal is jeopardized by faculty gender bias.

We discussed this work with Jo Handelsman, senior author on this gender bias paper, on episode #48 of the science show *This Week in Microbiology*. You can find TWiM #48 here (bit.ly/1LAU1sZ).

This article first appeared in the Virology Blog (bit.ly/1N6tJ4c) and is reprinted by the author's gracious permission.

Vincent Racaniello is the Higgins Professor in the Department of Microbiology and Immunology at Columbia University.

References

Moss-Racusin CA, Dovidio JF, Brescoll VL, Graham MJ, Handelsman J. 2012. Science faculty's subtle gender biases favor male students. *Proceedings of the National Academy of Sciences of the United States of America,* **109**:41, 16474-9

February 14, 2013

bit.ly/1Gnp7Ff

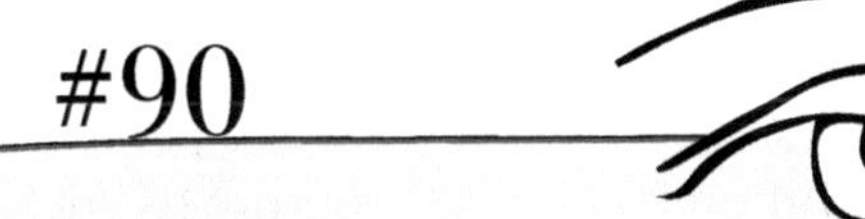

#90

by Welkin Johnson

Is it conceivable that, in evolution, a virus could switch from one genome type and replication style to another?

September 20, 2012

bit.ly/1hOOcgN

42

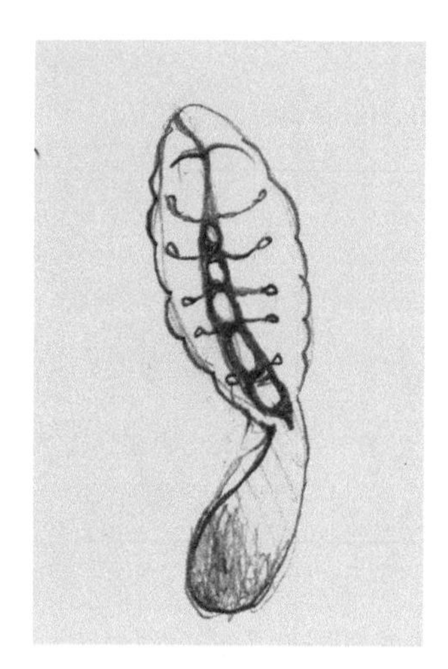

The Excitement of Clinical Microbiology

by Elio

Clinical microbiology, one of the major branches of microbiology, goes largely unnoticed by academics, in part perhaps because the diagnostic activities of microbiologists are pursued separately, in hospital and commercial labs. I'd venture to guess that many academic microbiology researchers have never set foot in one of these labs. Their loss, as it would be an exciting and rewarding experience.

This rift casts aside the historical roots of our field. At its beginning, soon after bacteria were found to be the cause of infection, there was a huge rush to develop methods for determining the identity of agents responsible for a patient's illness. Ingenious laboratory media were designed to favor the growth of certain organisms and to reveal their distinguishing properties, and serological techniques were developed alongside. In those days, there was no significant distinction between diagnostic microbiology and the rest. In time, as research in other fields of microbiology matured, these diverged more and more from this point of origin.

I find it unfortunate that such a cleft exists. From a practical standpoint, microbiology laboratories continue to make most of the diagnoses of infectious diseases in hospitalized patients and thus play a key role in medicine and public health. Further, the world of clinical microbiology can teach the rest of us a great deal. Some of you might recall the TWiM podcast (bit.ly/1LXN3MD)recorded at the General Meeting of ASM in 2011, where Andreas Baümler lifted the curtain on how tetrathionate broth, an old medium used for the enrichment of salmonellae in stool specimens, selectively enriches for these species. *Salmonella* carries out anaerobic respiration using tetrathionate generated in the colon and thereby outcompetes the normal flora.

Clinical microbiologists uncover new and important pathogens, perform the role of sentinels to alert of possible upcoming epidemics, provide statistical and clinical information regarding the pathogens currently on the scene, and spur demands on research to create novel diagnostic tools. In fact, the development of such tools is taking place so swiftly that in not too many years the practice of clinical microbiology may well become unrecognizable. Not only will the introduction and use of nucleic acid-based techniques continue apace, but other sophisticated techniques such as mass spectrometry will make diagnoses ever more rapid and accurate.

Let me back up a moment and share with you a brief nostalgic trip I took recently. I spent a morning in the clinical microbiology lab of the VA Medical Center in San Diego, just following the people around in their normal course of work and asking a bunch of questions. The nostalgia arose because this experience took me back to my early years. As a youngster, I had the good fortune to work in a pharmaceutical company in Quito, Ecuador (for an account, go to bit.ly/1MSCcDn, and see chapter 8). Since this was then the only facility equipped for microbiological work, it also served as the diagnostic lab for the local physicians. For a couple of years, I was more or less in charge of doing this work. And I loved it. To a kid in his late teens, what could be more exciting than finding out that a specimen contained such famous bacteria as staphylococci, typhoid bacilli, pneumococci, or shigellae? I remember the exquisite pleasure it gave me to isolate a somewhat unusual and difficult-to-cultivate organism, the agent of chancroid, *Haemophilus ducreyi*. I also isolated a strain of *Klebsiella pneumoniae var. rhinoscleromatis* that led to my first publication and which made it into the ATCC. It's still there. I felt that I had developed an intimate and personal relationship with all these exciting bugs. With such memories in my head, I stepped into the San Diego VA laboratory to learn, among other things, if anything has been preserved from the old days.

Sure enough, much of what is going on in this lab is now automated. Biochemical tests for sugar fermentations and other physiological attributes, once a test tube business, are now performed several dozen tests at a time by inoculating a multisectored plastic plate. The readout is of course electronic and the identity of the bacteria revealed by comparison to a database. Blood samples are placed in a special incubator and monitored continuously for anything that grows and produces CO_2. PCR-based tests are used to detect gonococci, chlamydiae, and many viruses, with more such applications on the horizon. For some specialized purposes, the material is outsourced to a central laboratory facility.

For all the new gadgetry, most of the lab's transactions required old-time skills, and abundantly so. I'll give you an example: crucial to the work is "reading" the Petri dishes 24 hours after inoculation with a clinical specimen (sputum, feces, abscess aspirates, etc.). Each specimen would have been inoculated onto several selective and differential media designed to favor the growth of certain species and to disclose some of their characteristics (e. g., colonial morphology, sugar fermentations, hemolysis).

Though "reading" the plates may sound simple, it surely is not. Every set of plates presents a different challenge about what to do next with each bacterial colony, and it takes considerable experience to get it right. In fact, laboratory technologists (a title not to be confused with *laboratory technician*) go to school for five years but are not considered to be fully trained until they have had three more years of actual practice! I could see why it takes so long. The variety of information that can be gleaned from just looking at these dishes and smelling some of them seems endless. I could almost see the wheels turning in the brain of an experienced technologist, Tracy, as she navigated in her head through a set of decision trees. She had to choose among many paths, such as picking a colony for biochemical tests, making a smear for a Gram stain, carrying out antibiotic sensitivity tests, and so on. I could hardly believe how many judgments Tracy made in front of me in a matter of seconds, all while thoughtfully answering my questions.

One key problem confronting the technologist is to tell if the bacteria growing on the plates are relevant to the diagnosis or are adventitious contaminants. Not always easy. A specimen of sputum may have been badly tainted by the contents of the pharynx, a urine sample may come from a "bad catch," or bacteria may have been picked up from the skin while drawing blood. Making this judgment may require looking at a Gram-stained smear of the specimen as well as at the colonies that grew from it. Jasmine, one of the newer technologists, showed me how looking at the smear helps her decide how to proceed. For example, if characteristic epithelial cells are present and neutrophils are absent in a sputum specimen, it probably contains mainly material from the throat and not deeper in the respiratory tract. And always there is the challenge of distinguishing potential pathogens from normal flora members, especially in samples that normally contain a large number of different organisms, such as throat swabs and feces. In addition, today's technologist runs PCR and immunodiagnostic tests, thus must be familiar with such technology and keep up with its continual advances. Proficiency with other special techniques is required to diagnose fungal and animal parasites. Such an impressive array of essential skills would be hard to duplicate with machines.

Not surprisingly, basic science continually interfaces with both the established and the newer techniques. As Josh Fierer, my friend, infectious disease physician, and microbiologist, explained to interns and research fellows, it is important to glean as much information as possible from microscopic examinations of Gram stains. Here, basic microbiology knowledge often enters the picture. For instance, an experienced person can often tell a-hemolytic strep from b-hemolytic ones by just looking at them in a smear. The former add new cell wall material from the septum outward, thus will look elongated just after they divide. The b-hemolytics, on the other hand, make this material mainly at the septum, so, after division, they eventually become spherical. And there are many such tricks. Amazing, how much information of great value can be gathered from simple procedures combined with keen observation.

Some of the challenges these labs face are distinctly new. From time to time, public health agencies tasked with detecting bioterrorism send them "unknowns" and ask them to determine the identity of the bacteria. Depending on the organism, this can be a challenge or it can be relatively straightforward. But either way, they better get it right. Juan, the technical specialist in the lab, allowed that having to identify these "unknowns" is a good thing for the lab because "*it keeps us on our toes!*"

I was curious about how the people in the lab felt about the possible impacts of future changes. All the signs point to continuing technological developments geared toward increasing the speed of diagnosis but likely displacing laboratory personnel in the process. But when talking to Monica, the supervisor of the lab, I appreciated her assurance that even in a very high-tech world, human skills will still be required. She felt that though technologies will change, the need for astute and experienced people will continue, even if the particular complement of skills required will somehow shift. She sounded confident. I was mighty glad.

November 05, 2012
bit.ly/1PvZl4E

#4

by Elio

What if all the bacteria and archaea on Earth decided to go on strike and stop their metabolism all at once? Which of the global cycles of matter would be affected first? How long would it take for life as we know it to come to a stop?

December 19, 2006

bit.ly/1GQ7DMK

43

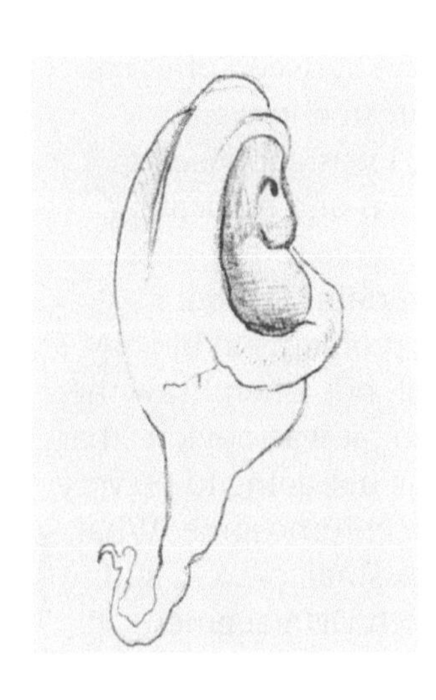

The Bacterial Chromosome: A Physical Biologist's Apology. A Perspective

by Suckjoon Jun

I entered the bacterial chromosome field in 2004 as a fresh Ph.D. trained in theoretical physics. Ten years is not long enough for one to gain the depth and breadth of a scientific discipline of long history, certainly not for an early career scientist to write an essay of the status of *A Mathematician's Apology* (Hardy 1940). Nevertheless, I agreed to write this perspective as a physicist who entered biology, because my colleagues are often curious to know what drives physicists to become (physical) biologists, and makes them stay in biology despite many challenges. I also wanted to share several lessons I have learned because, while some of them are personal and specific to my field, I have a good reason to believe that they might resonate with many future travelers. This perspective is for them.

I would like to start with the story of one of the most familiar and yet mysterious forces in nature—*gravity*. Galileo is said to have dropped two balls of different masses from the leaning Tower of Pisa in Italy some five hundred years ago. His experiment was to demonstrate that, on the contrary to Aristotle's *theory*, the falling rate of the balls was independent of their mass. A modern version of this experiment was performed on the Moon by the Commander of Apollo 15 with a hammer and a feather. When released from the same height at the same time, the two falling bodies hit the surface of the Moon simultaneously! On the Earth, however, the feather would have fluttered, as if alive, because of the air.

Initially attracted to the beauty of Amsterdam, I started my post-doctoral research at AMOLF, an interdisciplinary research institute known for exciting interactions at the interface between physical and biological sciences. The forces I was interested in were much less tangible than gravity. In particular, I was supposed to explain the driving force underlying segregation of a replicating chromosome in *Escherichia coli*. It sounded simple to me, except that I barely knew anything about bacteria, certainly without realizing that it was one of the long-standing problems in biology. I knew the DNA biophysics literature fairly well, but when I saw the beautiful 1992 illustration of *E. coli* in Goodsell, it was obvious that something like the wormlike chain model was not going to be very useful to understand segregation of the *whole* chromosome. What worried me was the directionality—if I were a small protein sitting on a replicating chromosome, could I tell which DNA segment belongs to which sister DNA? Physicists like questions like that, whether they are rooted in physics or biology.

The first challenge I faced was something unexpected. It is the aspect of "natural history" in biology. In physics, papers are relatively straightforward to judge their validity and importance. Physicists find the biology literature much less so, and they can be taken aback when they learn that the published data they modeled are later proven wrong. It can take a long time for physicists to learn which biologists are more reliable than others and who are more careful about what they choose to say about their data. This natural history aspect of biological research is something one cannot learn from the literature, as it is more about experience than intelligence. If the physicist learns biology from another physicist, there is a danger of learning a language and culture from a non-native speaker. Alternatively, if he goes to a biology conference to learn (assuming he already knows which conference to go to), typically the biology is overwhelming even for an experienced biophysicist. The best solution is to find reliable and generous biologists colleagues, who are not only willing to walk the physicist through the biology in an honest manner, but also genuinely interested in thinking about the problem from a different angle with the physicist. In this regard, I have been extremely fortunate to have worked with many generous biologists in the field, in particular, Stuart Austin, Conrad Woldringh, and Andrew Wright.

The important biology I learned early on from Conrad Woldringh regarded the physiology and cell cycle of *E. coli*. Conrad was particularly fond of studies in the 1950s and 1960s, such as the ones from the Copenhagen School of Bacterial Physiology (Maaløe and Kjeldgaard, 1966). Consider, for instance, *E. coli* growing in nutrient-rich medium, whose generation time is shorter than the duration of DNA replication. Since there must be a strict one-to-one correspondence between replication initiation and cell division (Mitchison 1971), a new round of cell cycle must start before the previous round of cell cycle is completed. Figure 1 illustrates a schematic development of overlapping cell cycles in *E. coli*. Why is this important? For one thing, we can entertain a *gedankenexperiment* about the role of hypothetical mitotic spindles in bacteria as shown in Figure 1. Under multifork replication, the hypothetical spindles that would pull two pairs of centromeres would mix the sister chromosomes, rather than separate them! From this "spindle paradox," I learned my first scientific lesson in the bacterial chromosome. That is, to really understand chromosome segregation, one should pay attention to the global, physical properties of the chromosome itself.

The polymer model of the bacterial chromosome Bela Mulder and I proposed in 2006 (Jun and Mulder, 2006) (1.usa.gov/1k8Bb2X) was based on the first lesson described above. Our two main ideas were that (1) when the right physical conditions are met, polymers with excluded-volume interactions will spontaneously unmix with each other even in a strongly confined space such as *E. coli*'s cellular volume, and (2) DNA will move much faster and more freely in the cytoplasmic space than inside the nucleoid, i.e., the chromosome dynamics should be considered in 3D, not in 1D. Some of the data we had modeled soon had to be updated by new data (Wang et al. 2006; Nielsen et al. 2006) (1.usa.gov/1LXNIxJ; 1.usa.gov/1jQ7LqT), but these two ideas were robust to the specific features of the data. However, an important question remained: what are the right physical conditions for spontaneous segregation? It took several years to answer the question in a series of theory papers (summarized in Jun and Wright, 2010 [1.usa.gov/1XxV02l]).

The next important lesson was the result of an identity crisis. At one point after having worked out most of the initial chromosome modeling, I saw real *E. coli* for the first time under the microscope in Conrad's lab across the street. They were very tiny, and they were so alive. Watching the tiny cells swimming and tumbling with my own eyes was a humbling experience; it made me realize I had no real understanding of what is meant by *living* organisms. Eventually,

I understood the fundamental difference between biological physics and biology, that biology in its heart is an experimental science and that, in biology, details—"prefactors" as opposed to "exponents"—really matter. I believe virtually all theoretical biophysicists go through a period of a similar crisis, and ask themselves how deep they should go into the biology. In my case, I came to a conclusion that I really wanted to understand *E. coli*, and it was only with a great fortune that I was given the freedom and resources to pursue my experimental ideas while a Bauer Fellow at Harvard (Pelletier et al. 2012) (1.usa.gov/1RdtVOj). For any physicist readers wanting to understand the biological reality and why their biologist colleagues think the way they do, I highly recommend to ask them to admit to their lab, at least for

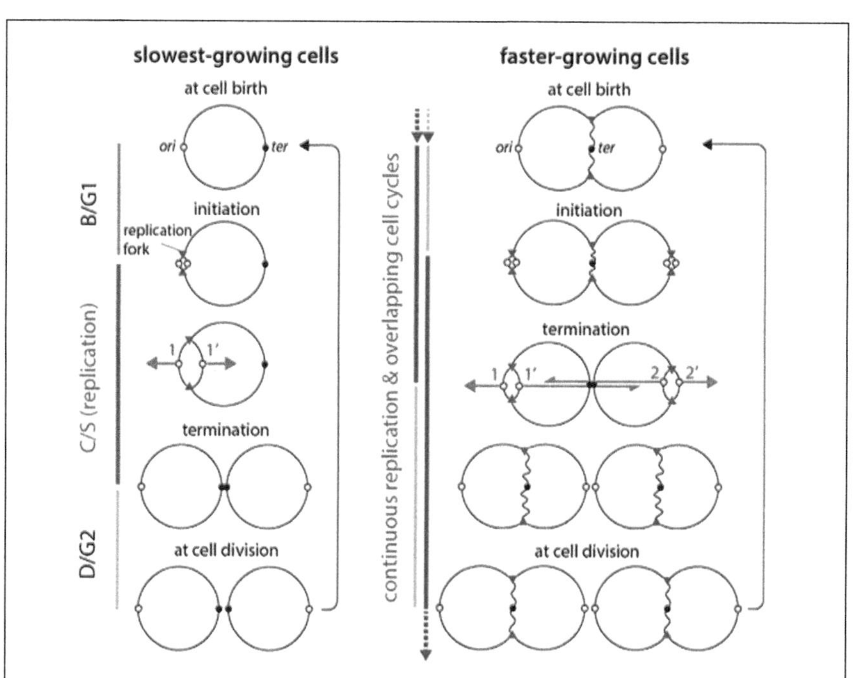

Fig. 1. The spindle paradox in *E. coli*. At the slowest growth rates, the chromosome rests in B/G1 before the replication period C/S starts. This is as in eukaryotes, and in principle hypothetical spindles can separate the duplicating *ori* regions (1 and 1′) to the opposite poles of the *E. coli* cell (red arrows). At faster growth rates, however, replication is continuous and multiple rounds of replication cycles overlap. The illustration on the right shows four copies of duplicates Ori's. The hypothetical spindles in this case will make the *ori* 1′ and *ori* 2 swap their positions. That is, the spindles would eventually mix the chromosomes, rather than separate them.

Adapted from: **Youngren B, Nielsen HJ, Jun S, Austin S.** 2014. The multifork *Escherichia coli* chromosome is a self-duplicating and self-segregating thermodynamic ring polymer. *Genes & Development* **28**:71-84.

one summer. The experience will change their view permanently.

I described at the beginning why the general mechanism of chromosome segregation and organization should account for all growth and cell-cycle conditions, and that it must come from the chromosome itself, not from the spindles. The final and most recent lesson is indeed about the *E. coli* chromosome during its general, overlapping cell cycles in living cells. This is due to the heroic effort of Stuart Austin (Youngren et al. 2014) (1.usa.gov/1LBcwOn). In 2006, two laboratories led by Stuart Austin and David Sherratt published important results on the organization of the *E. coli* chromosome under slowest-growing conditions, under which the cell cycles do not overlap (Wanget al. 2006; Nielsen et al. 2006) (1.usa.gov/1LBcwOn; http://1.usa.gov/1jQ7LqT). The two papers reported that, unlike the previous belief in the field, the left (L) and right (R) arms of the *E. coli* chromosome occupy each cell halves along the long-axis of the cell, such that the replicated chromosomes are organized as LRLR or LRRL. Stuart, however, did not stop there. His group started tackling what had been considered impossible. They started the measurements and analysis of intracellular positions of a couple dozen dual genomic loci markers under overlapping cell-cycle conditions. This is a daunting task because, at any given moment, every cell contains several homologous copies of each genomic locus, and deciphering the organization and dynamics of the whole chromosome based on their positional information was something that had never been done before. Nevertheless, Stuart's group relentlessly pushed their efforts without publishing anything for several years. When the task was done, the end result was simple, elegant, and surprising. The organization of the chromosome during multifork replication was that of a simple branched donut, with the two arms of the chromosome occupying each cell half along the radial axis of the cell. This result could not have been predicted based on our knowledge of slowest-growing cells, but it encompasses all the previously known results and moves beyond them into something new and general (Youngren et al. 2014) (1.usa.gov/1LBcwOn).

So what then does gravity have to do with the bacterial chromosome? I think it's remarkable that we accept the existence of an invisible, attractive force from the Earth, even when a sheet of paper flutters or even flies away in the wind. Perhaps it's the tangibility of gravity that helps us feel the concept. To me, physical forces intrinsic to the polymeric nature of the chromosome are as mysterious and yet tangible as gravity. Molecules, be they "air molecules" or proteins, are important in understanding the details of the motions. However, understanding the major physical driving forces, and how physics and biology have worked together during the course of evolution, will only deepen our appreciation of the role of the molecules. Recent work by Wiggins's group shows an exemplary approach towards such a direction (Kuwada et al. 2013) (1.usa.gov/1LXPrTD). As for me, I am convinced that the (in) tangible physical forces drive chromosome segregation no matter what, and physicists will continue to enter biology until one day the boundary between the two disciplines disappears.

Suckjoon Jun is an Assistant Professor in the Department of Physics, Section of Molecular Biology at the University of California, San Diego.

References

Goodsell DS. 1992. *The Machinery of Life*, Springer, Berlin
Hardy GH. 1940. *A Mathematician's Apology*, Cambridge University Press, Cambridge, UK.

Maaloe O, Kjeldgaard NE. 1966. *Control of macromolecular synthesis: A study of DNA, RNA, and protein synthesis in bacteria*, W. A. Benjamin.

Mitchison JM. 1971. *The Biology of Cell Cycle*, Cambridge University Press, Cambridge, UK.

March 17, 2014

bit.ly/1QNREV0

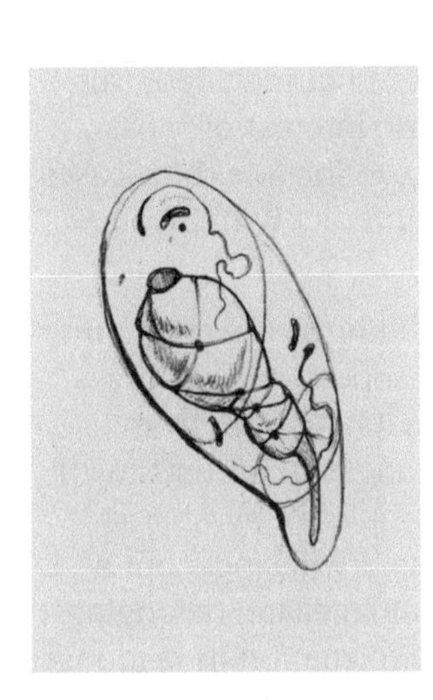

44

Whose Planet Is It Anyway?

by Elio

This is the title my friend Fred Neidhardt recently used for a talk, and a good question it is. I suppose that most microbiologists and the readers of this blog would split the answer down the middle, the biomass of this planet and the chemical transactions therein being about half microbial, half everything else. However, it's safe to say that most people, many scientists included, are unaware of the colossal importance of the microbial half, not only in biology and medicine but in geology, meteorology, and in our Earth's habitability. This state of affairs should not be unexpected, given that we have only became aware of much of this during the last few decades. I lived roughly the first half of my life carrying only a vague notion of the global importance of the microbial world. But now we know, and the word needs to go out. A measure of microbial literacy is required for anyone to understand the workings of our living planet.

Through the years, many influential writers have endeavored to convey the global influence of microbes to scientists and non-scientists alike. We can now add to these efforts a new contribution that speaks to scientists of all spheres, but especially to other biologists. It was recently published as a *Perspective* (1.usa.gov/1PPZqQM) in PNAS, a most appropriate venue. Entitled *Animals in a bacterial world, a new imperative for the life sciences*, it is authored by 26 scientists whose names are bracketed by those of Margaret McFall-Ngai and Jennifer Wernergreen. It deals specifically with the role of microbes in the lives of animals. While interactions with plants and the inanimate environment are not included, this seems a fitting focus given the anthropocentric interest of most readers. The other stories are for another day, to include the viruses, the most numerous of all players and which interact with all other living things.

Perhaps you think that in this blog I am preaching to the choir, and admittedly I am. But the sermon provided is superb and deserves sharing. The authors traverse the whole animal kingdom, from humans to fruit flies to shrimp to squid to sponges to protists, with stops in between, offering up fascinating examples of symbiotic relationships, some well known, others less so. I will single out a few that especially caught my fancy. A choanoflagellate, belonging to the oldest of animal phyla, initiates colony formation in response to bacterial signals. (might this speak of how multicellularity arose?). There is a shrimp whose embryos are protected from a fungus by the 2,3-indolinediole made by *Alteromonas* bacteria on board. A *Bacteroides* found in the guts of Japanese persons has genes for degrading the seaweeds in their diet, probably acquired from a marine bacterium. An alga protects itself by producing compounds (furanones) that mimic quorum-signaling molecules, thereby blocking communications between invading bacteria. And so on. But don't think for a minute that this is just a parade of astonishing stories. The authors use each example to elucidate a particular category of interactions, thus infusing the work with deeper meaning. Reading this article will sharpen one's insight. For those enticed to follow up any story, the authors provide a list of well-chosen citations in the supplemental material.

I could stop here, but can't resist listing other exciting material found here that I should have known but didn't.

- The fossil record tells us that early animals (in the Ediacaran) grazed on dense bacterial assemblages and that burrowing animals originated together with bacterial mats.

- A majority of the genes that animals carry descend from bacteria and archaea, or protists.

- Human-associated bacteria exchange genes at a rate that is 25-fold higher than those not living in host-associated host-environments.

- Up to one third of an animal's small molecules in the blood (its metabolome) are of microbial origin.
- Segmented filamentous bacteria with a reduced genome that live in the gut of some mammals are critical for the maturation of the immune system.
- At the bottom of the sea, bacteria feeding on carcasses make noxious metabolites that repel animal scavengers 10,000 times their size.

The authors pose some great questions: How have bacteria facilitated the origin and evolution of animals? How do animals and

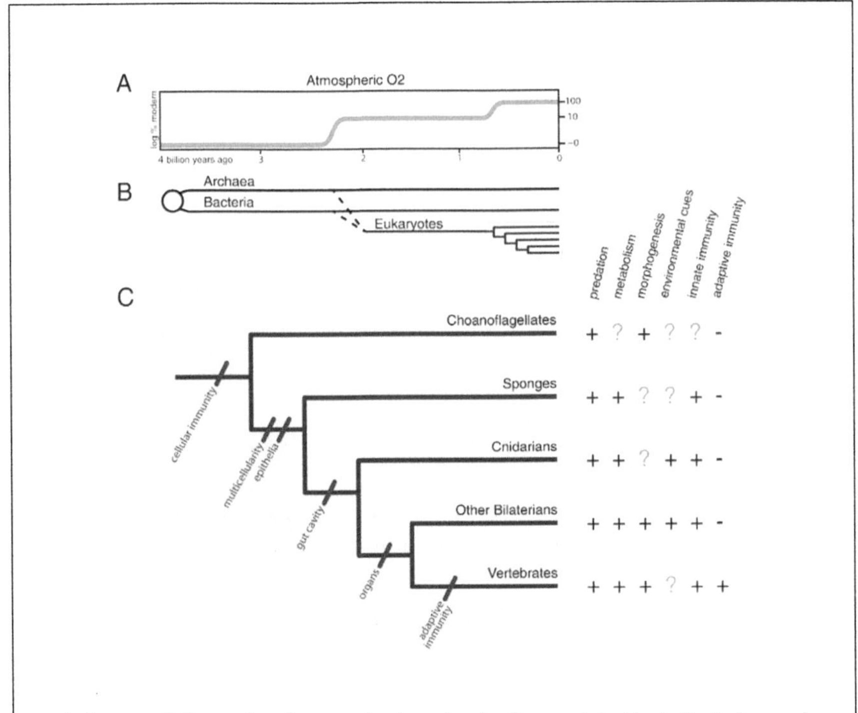

A phylogeny of choanoflagellates and selected animals, annotated to indicate the evolution of characters particularly relevant to interactions with bacteria. (Right) Interactions between bacteria and eukaryotes, corresponding to the phylogeny. Bacteria are prey, sources of metabolites, inducers of development in symbiosis (morphogenesis) and in larval settlement (environmental cues), and activators of immune systems.

Source. **McFall-Ngai M, Hadfield MG, Bosch TC, Carey HV, Domazet-Lošo T, Douglas AE, Dubilier N, Eberl G, Fukami T, Gilbert SF.** 2013. Animals in a bacterial world, a new imperative for the life sciences. *Proceedings of the National Academy of Sciences* **110**:3229-3236.

bacteria affect each other's genomes? How does normal animal development depend on bacterial partners? How is homeostasis maintained between animals and their symbionts? How can ecological approaches deepen our understanding of the multiple levels of animal–bacterial interaction? They point out that the answers to these questions are relevant to all biologists and beyond.

Progress is undoubtedly being made along the "it's time to accept that microbes are a big deal" front. The huge effort designed to understand the human microbiome and how it changes between people in different states of health and disease has certainly caught the public's eye. With the promise of personalized medicine, the microbiome of each person will be part of everyone's medical history. As the Romans said: *Suum cuique* (to each his own). The Earth Microbiome Project will follow, although where it will take us is for the future to tell us. We must thank this happy state of affairs in good part to the development of the various "omics." (By the way, if you find the term –omics grating, what do you think of subsuming them under the term *polyomics*?)

But for now, the task of changing the general perception of microbes in the scheme of things is still ongoing. Voicing a concern is one thing; delivering the message with such a formidable voice is another. I am grateful to the authors who have made such a commendable effort and thereby have contributed to both our understanding and our enjoyment.

Reference

McFall-Ngai M, Hadfield M, Bosch T, Carey H, Domazet-Loso T, Douglas A, Dubilier N, Eberl G, Fukami T, Gilbert S, Hentschel U, King N, Kjelleberg S, Knoll A, Kremer N, Mazmanian S, Metcalf J, Nealson K, Pierce N, Rawls J, Reid A, Ruby E, Rumpho M, Sanders J, Tautz D, Wernegreen J. 2013. Animals in a bacterial world, a new imperative for the life sciences *Proceedings of the National Academy of Sciences,* **110**(9): 3229-3236.

April 15, 2013

bit.ly/1MACL4R

45

If Microbes Could Tweet...

by Daniel P. Haeusser

Organized by Kara Schoenemann, Belkys Sánchez, and Minseon Kim. Tweets by members of the University of Texas at Houston MMG Department.

Each spring the Department of Microbiology and Molecular Genetics at UT Houston goes on our annual departmental retreat to Camp Allen for science, recreation, and festivities. After the first day of student and postdoc talks we have our dinner and a keynote speaker (this year we had a great talk from Tina Henkin from OSU).

Then, before the ever-popular "social time" that extends into the wee hours, the first-year students are responsible for running roughly an hour of "Science Fun Time." Typically, this consists of competitive games that may try and work in ways to embarrass professors, and always features a lot of laughing. This year they did a version of Pictionary with microbiology and molecular genetics themes.

The first years then finished "Science Fun Time" with a PowerPoint presentation of *If Microbes Could Tweet...* Prior to the retreat, they solicited department members to contribute humorous samples of Twitter messages that a microbe would compose were they capable of tweeting. The presentation displayed all the anonymous submitted entries. Some are a little esoteric, but many should be clear to anyone with some microbiology knowledge.

If you know all about Twitter, you can skip this paragraph and scroll right on down to the tweets! For those who may be unfamiliar, Twitter is a social networking site that allows users to post and read short 140-character messages (tweets). With such a short length limit these tweets succinctly convey a bit of news, a random thought, or a state of emotion at a particular moment in time. A hashtag (#) before a word or phase allows users to group and easily find posts of related topic or theme, and can often take a form of literary irony, or simple sarcasm. The @ sign is put before a username (handle) to identify and "tag" that user's account so they get notified the post is there. In addition to an official Twitter handle, users have a regular display name. Many thus choose to use their real name for the display, and something else for their handle (e.g., Robert Koch@PostulatePunk).

So now we want your ideas for *If Microbes Could Tweet...* Our department tended to pick microbes that we use in house as model organisms. But of course there are many more out there, perfect for some creative fun. So what would your favorite microbe use as its handle, and what would it tweet about its lifestyle and associates?

You can use this template to create your own tweets for your microbe(s) of choice. If you have your own Twitter account, tweet your ideas with attribution to the microbe and #microbialtweets, for example:
S. pombe #microbialtweets: Feels like I'm being torn in two... #headache #fissionproblems

P. Haeusser is an Assistant Professor at Canisius College and an Associate Blogger at STC.

May 13, 2015
bit.ly/1M2pUaX

46

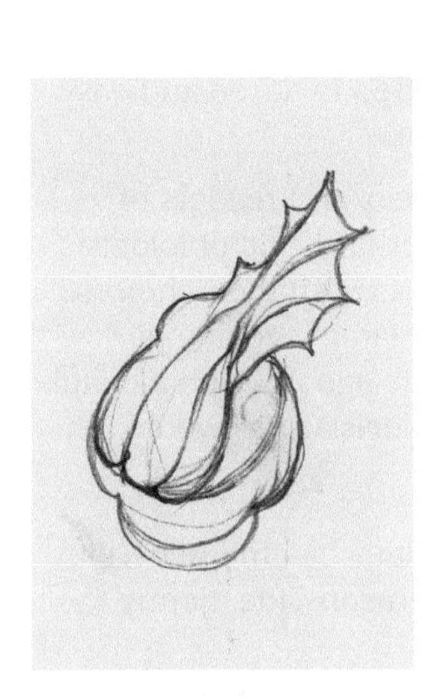

State Microbes

by Elio

Not long ago, I was interviewed twice by Michelle Norris of National Public Radio's *All Things Considered* regarding the burning issue of state microbes. The first interview was in response to the news that the state of Wisconsin's State Assembly passed a bill proclaiming *Lactococcus lactis* as its state microbe. I opined that this would be a most appropriate choice, given the role of this bacterium in making cheese, a matter of obvious importance to Wisconsinites. Alas, the state Senate did not take up the bill, so *L. lactis* will have to wait (perhaps in a lyophilized state) for future proposals. Michelle proposed that the other states now have the chance to be the first to adopt a state microbe. Listeners sent in their nominations, some of which we discussed during the second interview.

Mentioning the quest for state microbes to microbiologists and non-microbiologists alike usually results in a chuckle. Fair enough: in today's world, preoccupation with such ostensibly trivial matters may be frivolous. Still, there could be an educational point to this. Kids in each state, when presented with the state microbe, may want to figure out what makes the microbe special and even learn something about microbes in general. (Maybe adults, too!)

The interviews were good fun (the kinship in the names of our respective activities, STC and ATC, was a comfortable point of departure). Michelle was exceedingly friendly and helpful, likewise her off-the-air colleague Melissa Grey (author of *All Cakes Considered*). Both were eager to delve into matters microbial. Each time the interview lasted some 15 minutes, from which they selected the material that was aired. We chatted in a free and easy manner that I found both relaxed and pleasant.

Here is a list of nominations from NPR listeners and readers of *New Scientist*, as well as some of our own. Perhaps microbiologists in each state will find it worthwhile to come up with their choices. Those with political connections to state legislators should pay special heed. I omit suggestions of pathogenic agents (with a couple of exceptions), on the supposition that state legislatures would not take kindly to them.

- Alabama: *Karenia brevis*, a dinoflagellate, aka the Red Tide Alga. A shoo-in if you think of the Crimson Tide. (Jenny Ridings, *New Scientist*)

- Alaska: The permafrost bacterium *Carnobacterium pleistocenum* found in 35,000-year-old ice. (Rowan Hooper, news editor, *New Scientist*)

- Arizona: (1) *Thiobacillus ferrooxidans* for its role in copper leaching. Arizona is the top copper-producing state in the US. (Elio)

  (2) *Penicillum* for the penicillin it makes, to be used to cure gunshot wounds due to the recent illegal immigrant law. (NPR listener)

- California: (1) *Saccharomyces cerevisiae,* for its importance in making wine. (Elio) (Also suggested for several other states) (2) The city of Los Angeles got two nominations for a "city microbe": *Clostridium botulinum,* the source of botox. (NPR listener; Rowan Hooper, news editor, *New Scientist*)

- Florida: (1) Retirement communities in Florida would appreciate the 250-million-year-old Lazarus bacillus *Bacillus permians.* Note that there is far from universal agreement about the longevity of this bacterium. (Rowan Hooper, news editor, *New Scientist*) (2) The Sunshine State can share its ample sunlight and coastal waters with a photosynthetic marine cyanobacterium, e. g., *Synechococcus elongatus*. (Elio)

- District of Columbia: *Cupriavidus metallidurans* (formerly *Ralstonia metallidurans*), the gold mining bug that turns soluble gold into nuggets. They could use it there. (Elio)

- Hawaii: Another state claiming *Lactococcus lactis*, here fermenting taro into poi. (Merry)

- Indiana: *Zymomonas mobilis,* a bacterium that produces ethanol very efficiently. Indy 500 cars run on ethanol. (AmoebaMike, *New Scientist*)

- Iowa: *Bradyrhizobium japonicum*, a nitrogen-fixing symbiont of soybean plants. Iowa is one of the top soybean producers in the US. (Elio)

- Kansas: MRSA, for illustrating evolution in action. (several NPR listeners)

- New Jersey: (1) *Streptomyces griseus*, the bacterium that makes streptomycin, a pioneer antibiotic discovered at Rutgers University by Selman Waxsman. (Elio) (2) Sewage methanogenic bacteria, for New Jersey's famous marshland garbage dumps. (Rowan Hooper, news editor, *New Scientist*)

- New Mexico: The "indestructible bacterium" *Deinococcus radiodurans* that probably survived the Trinity A bomb test carried out in New Mexico in 1945. (Rowan Hooper, news editor, *New Scientist*)

- Nevada: Home of the neon glow in Las Vegas gets the flashing light of *Vibrio fischeri*. (Rowan Hooper, news editor, *New Scientist*)

- Rhode Island (1) *Epulopiscium fischelsoni*, the biggest known bacterium (never mind *Thiomargarita*) for the smallest state in the Union. (NPR listener) (2) The nanobacteria (Ed. note: assuming they exist). (Rowan Hooper, news editor, New Scientist)

- Texas: The oil-eating *Synthrophus* may be useful for cleanup of oil spills. (Rowan Hooper, news editor, *New Scientist*)

- Utah: The salt-loving *Haloarcula* for the Great Salt Lake. (Rowan Hooper, news editor, *New Scientist*)

- Virginia: The Epstein Barr virus or kissing bug because "Virginia is for lovers." (NPR listener)

- Washington: Here they may appreciate the rain-making bacterium *Pseudomonas syringae*. (Rowan Hooper, news editor, *New Scientist*)
- Wisconsin: *Lactococcus lactis*, the essential cheese-maker. (Wisconsin State Assembly)
- Wyoming: *Thermus aquaticus*, which was isolated from a Yellowstone hot spring and went on to make a great living as the source of the Taq polymerase. (Elio)

Can you think of others?

May 6, 2010

bit.ly/1jy7EjH

#81

by Elio

Given that most mutations are deleterious, why does the mutation rate not evolve to zero?

November 10, 2011

bit.ly/1MQxwT1

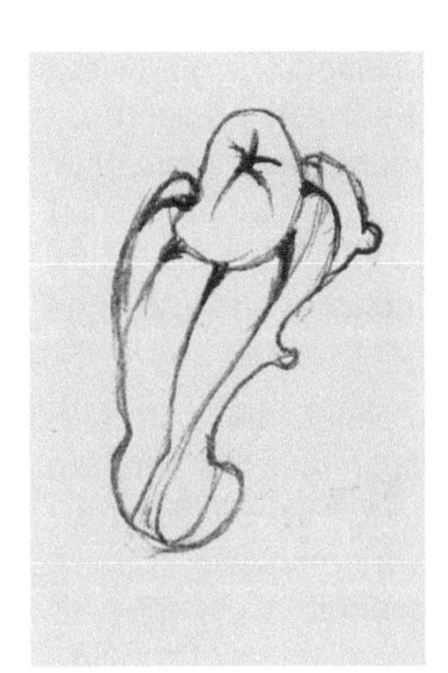

47

Recalling the Good in the Good Old Days

by Elio

In its early days, ca. 1945-1965, molecular biology was a particularly collegial undertaking, characterized by free sharing of research data and a relative lack of egotistical behavior. The reason for this marvel may well have been that there was so much to discover—so many low hanging fruits—that there was room for everybody and enough money for the pursuit. This was a special period in the history of science, not only because stunning discoveries were made with great frequency, but also for the way science was done.

Here is an example of selflessness that I witnessed in person. In 1961 I was a faculty member at the spanking new medical school of the University of Florida. These were not ordinary times. A period of just a few years saw the discovery of the semiconservative replication of DNA by Meselson and Stahl, the postulation of the operon model by Jacob and Monod, and the elucidation of the genetic code by Nirenberg and Matthaei. In a moment of hubris, our faculty invited the future Nobelist Salvador Luria to visit and give a seminar on his research. Luria was one of the founding fathers of this new science, having made stellar contributions to microbial genetics. His name is associated with the Luria-Delbruck experiment that established with simple clarity that mutations were not induced by the environment but rather occurred at random. Luria's acceptance of the invitation was a great moment for us.

We went as a delegation to our little airport and waited for the great man to come down the small airplane's flight of steps and sample the Florida air. Now, we knew that Luria was a person of strong views, not one to hold back his opinions, but we were scarcely prepared for his first pronouncement. On looking around, he blurted out, "*I hate palm trees!*" So there! Having swallowed that one, we proceeded directly to the auditorium where he was to talk. We had managed to assemble a large crowd for this presentation, including medical and other students. The spacious hall was packed.

Luria ascended to the podium, took out some papers and said something to this effect: "*I was going to tell about my work, but I just received a manuscript from Paris that deals with far more important results, so I will talk about that instead.*" The work he chose to talk about, done at the Pasteur Institute by Brenner, Jacob, and Meselson, provided definitive evidence for the existence of mRNA. I can't recall ever again witnessing that—a speaker deciding that because of its importance, someone else's results trumped talking about his own. But unusual as this action was, it seemed entirely in keeping with the spirit of the time. Can you imagine someone carrying out such a gesture nowadays?

October 24, 2013

bit.ly/1M2q2r1

48

The Attendee's Guide to Scientific Meetings

by Julian Davies

There are many aspects to the business of going to a scientific meeting. Different protocols are required for meetings of different sizes and classifications. A general meeting, such as the massive ASM or ASBC gatherings, attract many people with many justifications for attending; it is difficult to provide guidelines for meetings of this type unless you are an invited speaker or a poster presenter. In these latter cases, being as you are there as an invitee, you must make the effort to justify your invitation. You are there to perform, to present your scientific efforts to a larger audience for evaluation. As a privileged attendee, it is your responsibility to showcase your work in the best possible manner. If giving a talk, rehashing old PowerPoint slides simply will not do; your audience is expecting something new, interesting and stimulating (more on this to come later).

Attending one of the mammoth meetings, such as the ASM general meeting, year after year is common practice for many people—perhaps for you. This may be due to the fact that when you are a member of a committee or an Editorial Board, the annual meeting is a convenient place to meet and discuss why the impact factor of your journal is going down, when everyone else's is going up. However, there is another, perhaps more compelling reason to attend these large congregations, and that is to see old friends and to make new ones.

Going to see old friends—especially for the over 60s—is an exercise that is fraught with danger. You would be wise to plan your tactics accordingly. Why? At each year's meeting, you are one year older and your memory is one year more capricious. Will "old X" look the same? Has W got a new partner? Or an additional partner/friend? Has P retired or moved? (Insert names as appropriate.) For these and similar situations it is best to attend the opening reception cautiously.

For maximum satisfaction and poise, the following strategies are recommended. Wearing dark glasses can be a good ploy (if you can still see in a dimly lit room). It is also essential that you pin your name tag in a place where your full name is not readily visible—partly behind your lapel, for example. When you arrive at the reception, circle the outside of the room and pick out a few people that you do recognize and whose name you can recall. Stride boldly up and greet each of them in turn using his or her name. If you are of the same approaching-senior age group, there is a 50% chance that they will recognize you but not remember your name. Since they cannot see your whole name card, you will likely witness their desperate efforts to decipher your name. You will then begin to feel good! Once contact has been established, you can ask them if they remember the name of a person nearby.

However, sooner or later you will encounter someone you recognize but whose name escapes you. This requires a different tactical approach on your part, since he or she may know your name, despite your partially hidden name card. Try to approach this person surreptitiously, perhaps joining a group that is walking by and sneaking glances in the direction of your target's name card (dark glasses can help here). If you are unfortunate enough to actually come into chance contact with someone who addresses you by name, and you do not know his or her name, other gambits are needed. Assuming that theyhe or she is wearing a name badge, it is bad form to glare at it before responding. There are two good tactics to apply here. Give a show of recognition but then grab your handkerchief and explode with a fake sneeze ("*Good to see...atishoo*"); as you turn away politely this gives you the opportunity to read their name badge. Another ploy is to start to greet them and then quickly point and say something like "*Look! Isn't that Elio Schaechter with a beard over there?*" He or she will turn, giving you plenty of time to read their name.

Other variations can be tested, and you will soon become skilled and practiced in the art. You will then enjoy opening receptions in ways you had never expected!

Julian Davies is Professor Emeritus at the University of British Columbia, a Fellow of the Royal Society, and past president of the ASM.

December 3, 2009

bit.ly/1GQle6x

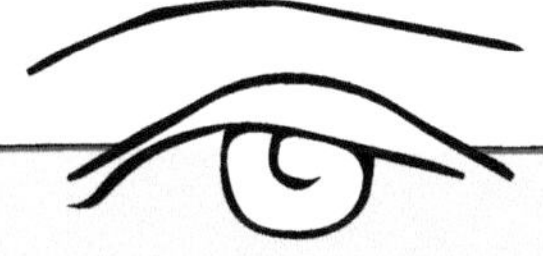

by Elio

Can you think of any group of living organisms that is not host to some virus?

March 24, 2007

bit.ly/1KlcGF9

PART 5

Personal Notes

Along the way
I could not resist
a few other personal
comments about
this business
of being a
microbiologist.

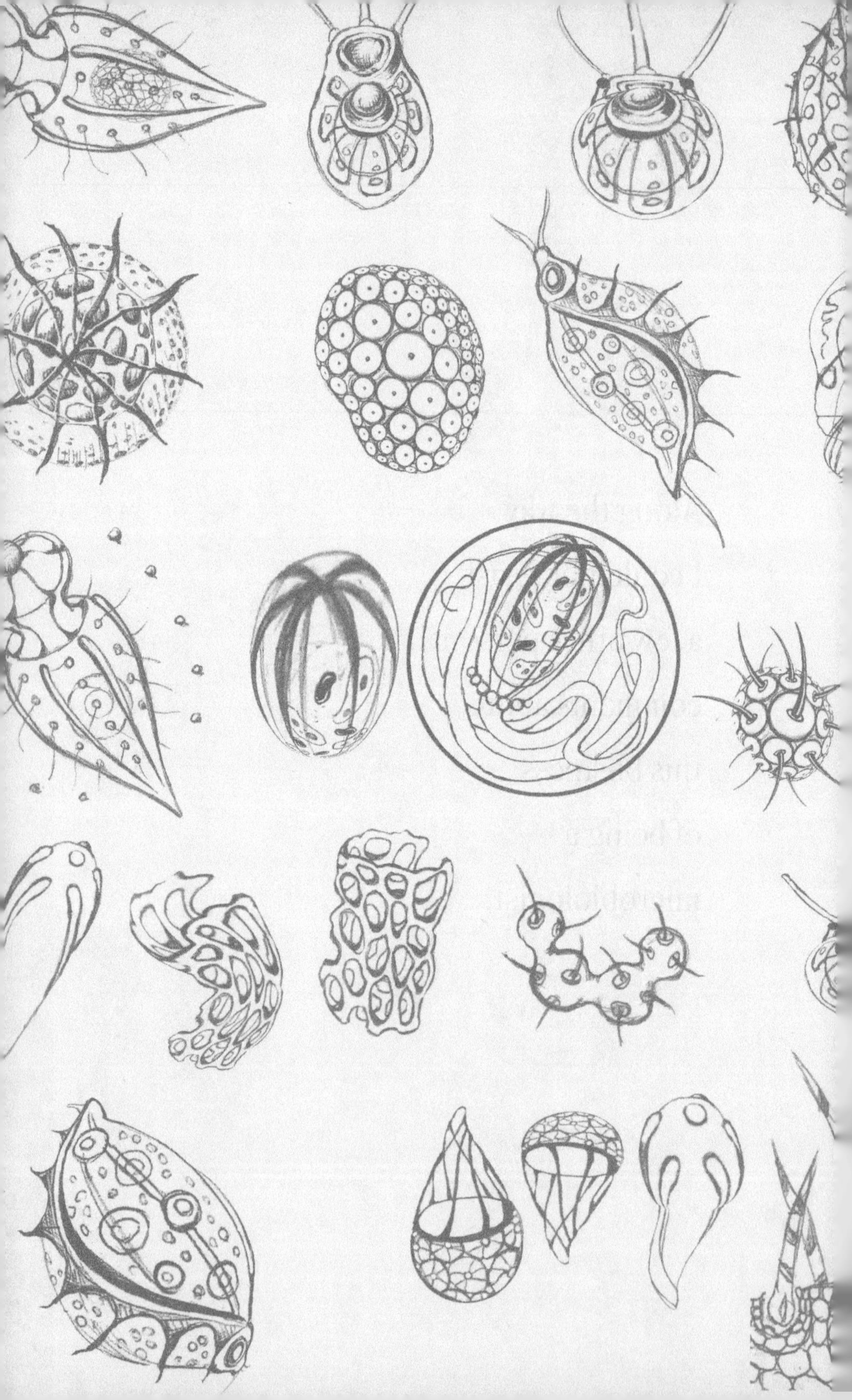

49

A Personal Note: The Encyclopedia and I

by Elio

For the last couple of years, during hours not spent blogging, I served as the editor-in-chief of the third edition of the *Encyclopedia of Microbiology*. If this sounds like a large undertaking, it is. The work consists of nearly 300 entries compiled into six volumes that weigh *in toto* close to 50 lbs! Fortunately for those with weak backs, this edition will also be available electronically. My contribution to this project was reduced to human dimensions because practically all the work was done by an editorial board consisting of some terrific microbiologists. They chose the authors and edited the manuscripts. To see for yourself how distinguished these people are, go to bit.ly/1XxWlGy. Did I critically read each entry from start to finish? Not on your life! (However, my curiosity was sufficiently piqued that I scanned—or more—each entry). Most of what I did was to cheer editors and authors along. I wasn't always good at this, but finally this Gargantuan creature emerged from its womb, hale and well.

At the outset, my job was not made any easier by the fact that I took it over from Joshua Lederberg. Enough said. On the other hand, he thought that an encyclopedia of microbiology was worth putting out. This answered *that* question. Who am I to argue? Still, let's bring it up one more time: Why amass a mountain of microbiological knowledge in this format? With Wikipedia and other Internet material, manuals and textbooks, review articles and research reports, is there not enough to satisfy every manner of thirst for knowledge? I agree, up to a point. There is still something unique to our effort: the depth of coverage is reasonably consistent, it's all together between covers—real or virtual—and, above all, the content endured serious quality control. To quote from Lederberg's preface to a previous edition:

> *"The reader wants access but not to be flooded with information and, of course, it's hard not to be inundated with material, even just that which is essential and important to one's scientific existence, not to mention everything else that comes to one's attention. We want quality assurance so we look for assistance, we look for filters, we look for peer review as a method of assisting us in navigating through the blizzard of available information that will only worsen."*

I am happy with the outcome. The entries were written by authorities in each field who did not stint in their efforts. Most of them are as good as I've seen. I am not advocating that you go out and purchase it. It's quite a lot of money and can only be afforded by those with unusually deep pockets. It's meant to be obtained by libraries, with a site license for the electronic version. I wouldn't mind if you ask your institutional librarian to get it (although it won't line my own pockets, as I have already been paid in a lump sum, rather than in royalties).. Also, my collaborators and I would be very happy if you were to provide us with some feedback. A review for this blog, maybe?

Do I have stories to tell about this exercise? Not many, which attests to how smoothly it went. A few authors came up with ingenious excuses for tardiness, the equivalent of "*the dog ate my homework*." But that was to be expected from busy people. And it all worked out in the end; the occasional setbacks turned out to matter little. One thing I had to keep in mind was that this was not a textbook. Thus, one of my minor but unforeseen tasks was to expunge words such as *Microbial* and *Bacterial* from the beginning of titles—an encyclopedia is organized alphabetically.

Here's a challenge to those audacious (or zealous) enough to read it from beginning to end: I will reward the effort with a bottle of good wine for the first credible claim. I'll set one aside now. (By the time anyone claims it, it will be well aged.)

The French encyclopedists of the eighteenth century Age of Enlightenment sought to catalog all human knowledge. Did we succeed in cataloging all microbiological knowledge? Hardly. But we may have made a dent and, we hope, may have captured some of the excitement in the field today.

March 26, 2009

bit.ly/1jy7KYt

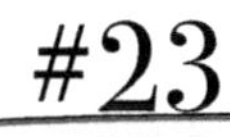

#23

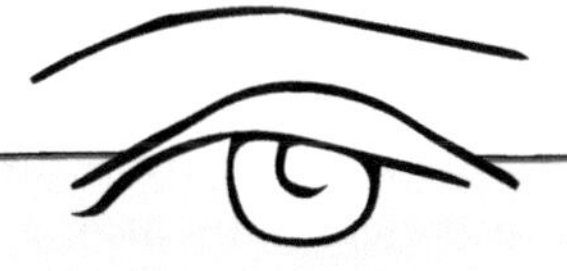

by David Lipson

Methanogenesis is characteristic of Archaea (and, according to some, plants as well). Why don't bacteria make methane?

November 1, 2007

bit.ly/1Nl4CZs

50

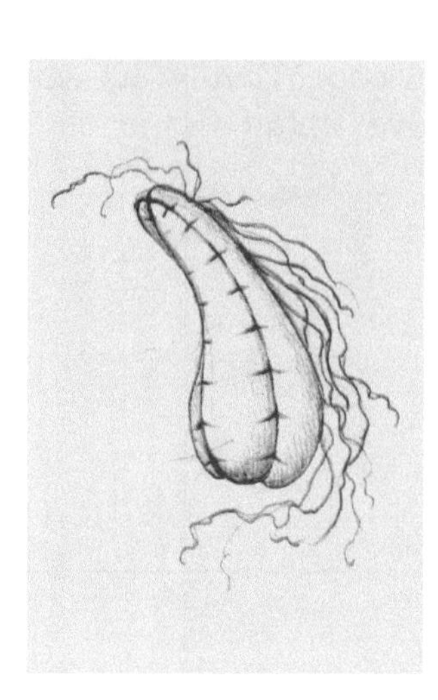

Musings: The Guild

by Elio

The dictionary defines a guild as "*an association of craftsmen in a particular trade.*" Let's extend this term a bit and apply it to the community of microbiologists. I make this point because I am intensely aware that I belong to a group of people united by their particular professional interests. Whenever I meet another microbiologist for the first time, I feel a high degree of affinity with him or her. We can immediately converse about common interests, and especially about people we both know. There is usually only one "degree of separation," two at most, depending in part on generational differences. How different this is from my encounters with members of other guilds, e.g., lawyers, carpenters, policemen, engineers.

Looking into this a bit deeper, I find that this sense of belonging matters a lot to me, and to my mental well-being, no less. I have derived and continue to derive emotional support from being a member of a community of people joined together by common experiences and shared concerns. You can now see why I consider my year as president of the ASM to have been one of the highlights of my professional life. The memory recedes, as the date was 1985, but I still treasure the experience.

This feeling has been deepened by the fact that all branches of our science have so much in common. Microbiology, as I and others have stated before, is a truly integrative science. The various branches converge in terms of both experimental approaches and conceptual foundation, which makes for ready affinity among its practitioners. I remember, going back perhaps 40 years, when I met a stranger on a train in Europe and learned that he, too, was a microbiologist. This was tempered, however, by his telling me that his field was dairy microbiology. I found this to be remote from my interests and, best I remember, I soon changed the topic of the conversation. This would not be the case today, as I would delight in the fact that this (and most other microbiological specialties) was just one topic in environmental microbiology, ergo microbial ecology, with a strong genomic component. Surely, there would be a lot that we could talk about.

What prompted these musing was the recent case of Bruce Ivins, the Fort Detrick microbiologist who took his life, apparently in connection with the protracted anthrax investigation. I didn't know Dr. Ivins and have had no connection with people working at labs such as that one. I have no way of knowing with certainty whether the allegations of his involvement with bioterrorism are founded or not. But whatever the case, I am deeply saddened by this event. He was a member of the guild, thus a fallen colleague.

September 4, 2008

bit.ly/1NRVEpM

51

The Magasanik Paradox

by Elio

The journal *Nature* published a study of the 100 most cited scientific papers that provides food for thought to the working stiffs (I used to be one). First of all, I find it quite natural for any researcher to keep an eye on how often his or her work is cited. Other than a personal reaction from colleagues and friends, there is no better way to find out how well one's paper is accepted beyond the confines of one's own lab. And who would not want to know? So, the topic of frequency of citations touches most of us.

The paper says that by far the most cited papers deal with techniques. The Lowry method (1951) for determining protein concentration tops the list with over 300,000 citations and is followed by a raft of others that include methods of biochemistry, bioinformatics, and phylogenomics. Neither Einstein's paper on the special theory of relativity nor Watson & Crick's discovery make the list of the top 100. The reason for this seems clear. Once an important discovery becomes part of the scientific canon of the time, there is no need to quote who did it. Woe if we did—every time we mention water, we would have to add: Lavoisier, A. 1783. *Observations sur la Physique*. 23:452 – 455.

But let me be a bit more nuanced. A factor that comes to mind is not just the number of citations, but their timing. This brings up what I will call the Magasanik Paradox, named after that wise bacterial physiologist, who is quoted as saying—more or less verbatim—that the more important a publication, the shorter the time it will be cited. This metric was not included in this report, although the data certainly exists and can be extracted for any individual paper. So, there is a bit of homeostasis-for-the-ego here: if your paper isn't all that important, at least it stands a chance of being occasionally cited for a long time!

January 28, 2015

bit.ly/1LAr3ah

by Elio

Many bacteria, e.g., *E. coli* and *B. subtilis*, regulate their gene expression via a large number of distinct devices that operate on almost all conceivable biochemical levels. Why so many?

March 24, 2008

bit.ly/1NRFnRD

PART 6

The Ways of Microbes

We didn't mean the blog to be about "heavy science." That can be found elsewhere. But some science has made it within its pages, and a good thing it is.

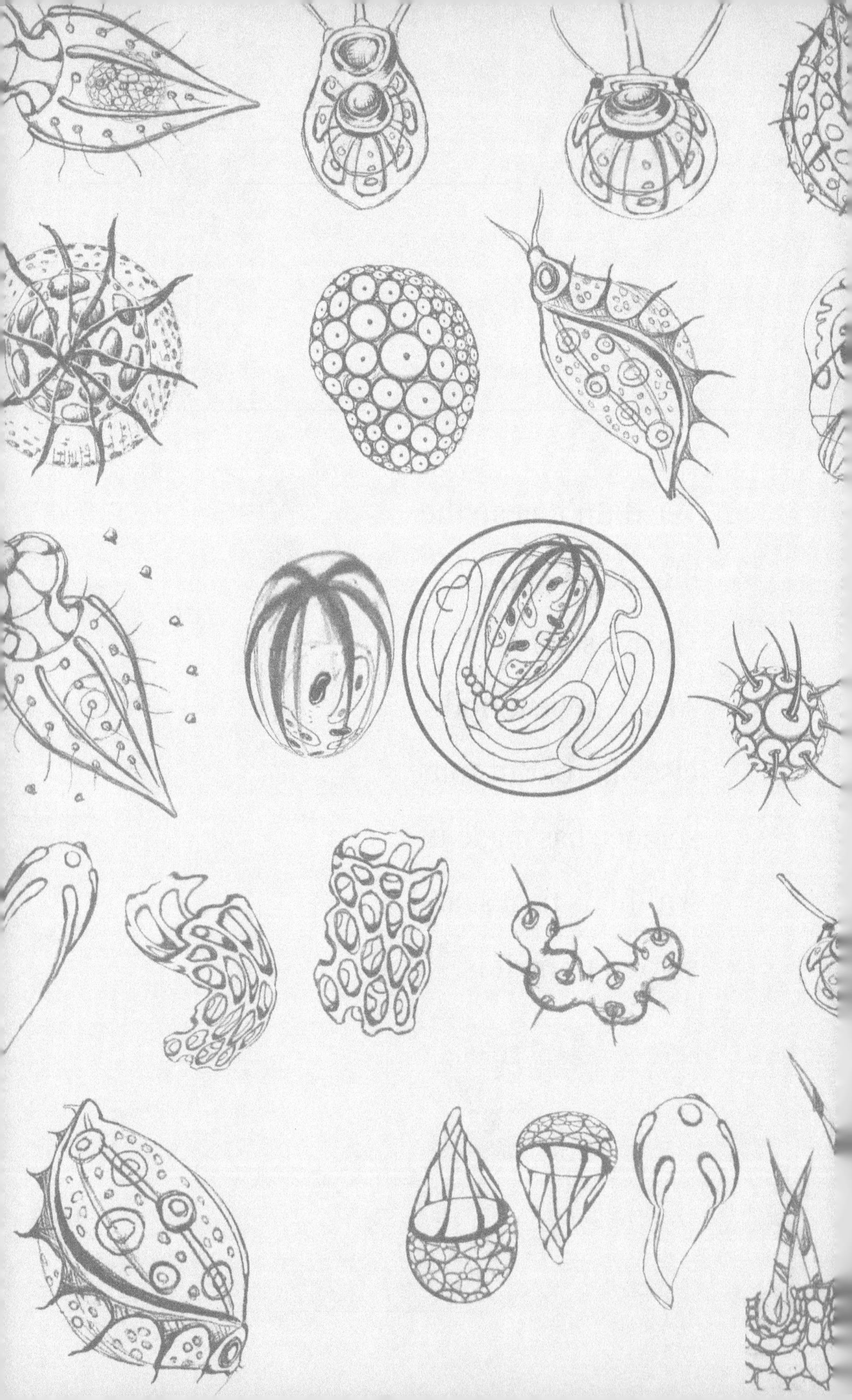

52

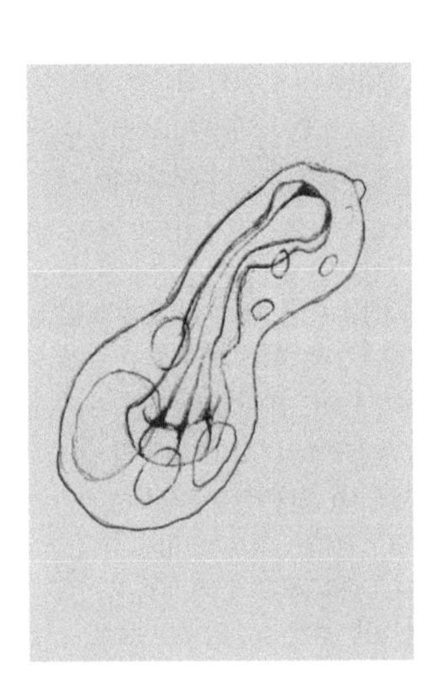

The Parvome

By Julian Davies

This is in response to Mark Martin's prompt*; he is an interesting and eclectic character. Well, I was not inclined to write anything, but since this week has been spent on grant writing (for not very much money, this being the style in Canada) and today, being a sunny day, I decided to cut the grass but the handle kept on coming detached from the push mower which put me in a good mood for a rant (not a blog); I am also upset about the fact that Roger Federer lost again and I don't like to see bad things happen to my heroes. But to come back to Mark's prompt, why do microbes make so many small molecules?

To start with I apologize for the fact that the question is biologically incorrect, since the only people who can answer "why" questions are priests. However, you know what I mean. I am very enthusiastic about the fact that the scientific world is becoming interested in biomes and their importance to eukaryotic life; it is about time! Jeff Gordon, Jeremy Nicholson, and others have confirmed the notion that bacteria are really good for us humans and we can begin to think seriously about microbes as being truly important living beings. As you know, I have long been interested in how microbial small molecules fit into this picture.

---

*Mark's prompt: "*I would love to see a post from Julian Davies from UBC over his idea (presented at ASM) that there exists a microverse of small molecules in nature which have profound effects on microbes in the natural world.*"

As I pointed out in an earlier letter, there is a humungous microbial world of small molecules of great structural diversity (Mark Martin suggested *parvome*, from the Latin, *parvus*, for "small," and I agree; I have always wanted my own "ome"). Microbes are the most accomplished chemists on earth; interestingly, one of the greatest challenges for organic chemists is to design synthetic approaches to these molecules and many naturally occurring compounds remain to be so made. But these small molecules were not produced to be PhD projects, and neither were they made to cure our infectious diseases. The latter is one of the best examples of anthropocentric thinking, in my opinion. We use bacterial small molecules as drugs precisely because they exhibit dose-response curves. At high doses they are good for us, but at low doses they must possess activities that are good for microbes, more especially for microbial communities (the only way that microbes exist in nature—well, maybe not in invasive infections).

So what do small molecules do? Our studies show that all natural products are bioactive, in the sense that at low (sub-inhibitory) concentrations they modulate bacterial transcription. Depending on the compound and its target, each small molecule influences a distinct spectrum of promoter responses. To me and some others, this implies that in the environment, bacterial communities are modulating their activities using a wide range of small molecule signals.

I could say that this was a form of homeostasis, but it is premature to make this conclusion. I could also say that bacteria all have their own cell phones, but this is a primitive anthropomorphism! I think we must accept that this is a different form of cell-cell signaling than is normally discussed, since the small molecule ligands interact with macromolecular receptors such as ribosomes, or the DNA replication, transcription, and cell wall complexes. All this remains to be proven, of course, and some good imaging studies in soils and other environments would help.

On the other hand, some small molecules are so toxic that it is difficult to see how they could be anything other than real antibiotics. A good example is the enediyne class produced by a number of different bacteria and used as antitumor agents. Bacteria make them, use them (for what purpose?), and survive! The self-protection mechanisms must be extremely effective since one slip means death. Incidentally, I wonder if the natural bacterial protection mechanisms could be used to control the activities of these potent compounds during cancer chemotherapy? Someone must have thought of this.

Finally, there has been much discussion of late about the "environmental resistome." One can certainly find endogenous antibiotic resistance genes in most bacterial populations by cloning and testing in the lab. But is this their real function in nature? Could resistance be due to pleiotropic effects? Is much of the clinically significant resistance simply due to the over-expression of a heterologous gene in a different cytoplasm?

So there you are. I hope that you have no concerns about my sanity; you can blame this post on Mark! Yours in wonder (about microbes),

Julian Davies is Professor Emeritus at the University of British Columbia, a Fellow of the Royal Society, and past president of the ASM.

September 1, 2008

bit.ly/1OPulel

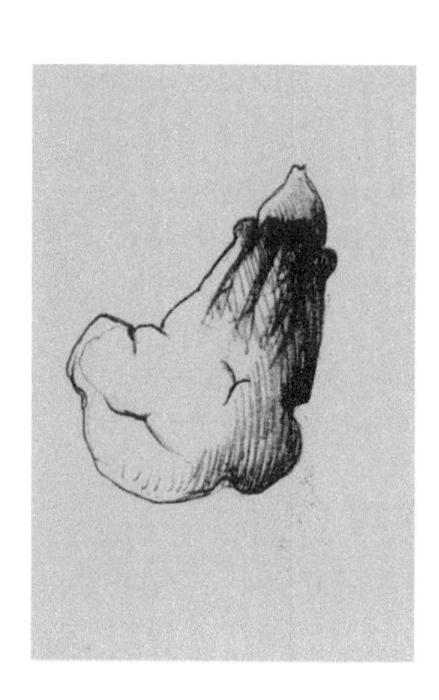

53

The World Is Pleiotropic

by Elio

So there! By this rash assertion I mean that in the biological world almost every macromolecular constituent is likely to function in more than just one way, i.e., is pleiotropic. One and the same protein may be enzymatic, regulatory, and structural; one nucleic acid, informational, structural, and regulatory; and so on. Clarence Jeffery called this "moonlighting."

Most biologists know this, but how many think it's a big deal? I do, based on little more than the belief that living things are parsimonious and make efficient use of their parts. Doesn't it seem reasonable that our own 30,000 or so genes (a puny number, one would think) each encode many functions?

Linc Sonenshein put it well in a recent e-mail:

> *"I believe that within 10 years, we will view all proteins as presumably multifunctional; that is, we will view monofunctional proteins as a rarity. It took us so long to come to this realization because we were raised on the one gene-one enzyme dogma and the great specificity of biological reactions, implying highly specialized functions that evolved to be the best they could be. It never occurred to us that cells (bacteria in particular) learned to do things that we hadn't thought of or that every protein is a compromise between the ideal for one of its functions and the less-than-ideal version that has the ability to carry out a second function."*

Now for something more specific. But first, a disclaimer: "function" can be taken in many ways, cutting across biochemical and physiological categories. I don't want to get into this here, thus will use "multifunctionality" in a broad sense. I hope that's OK with you.

A recent review (bit.ly/1RACD91) focuses on the multifunctionality of one particular set of proteins, those of the ribosomes. The numerous ribosomal proteins are juicy candidates. By definition, they are integral to the ribosomes and thus intimately associated with various RNAs and with translational activities. Evidence that they can also play a role in transcription has been accumulating for some time. Already by 2000, ribosomal proteins were front-and-center in a review (1.usa.gov/1MhCvY1) entitled *Proteins Shared by the Transcription and Translation Machinery*.

The multiple functions of one ribosomal protein, S1, are an old story in microbiology. It's been known since the mid-1970s that the RNA phage Qβ uses a replicase composed of four subunits, one of which is indeed ribosomal protein S1. Two others are also borrowed from its *E. coli* host: the protein-synthesis elongation factors EF-Tu and EF-Ts. Thus, only one of the four is phage-encoded. Protein S1 is a master of versatility. In *E. coli* it is required for ribosomes to recognize the translation initiation codon of most messenger RNAs; it stimulates a phage T4 endoribonuclease to inactivate some unneeded phage mRNAs by cleaving them in the middle of their Shine-Dalgarno sequence; and it seems to help out other ribonucleases, as well.

Other ribosomal proteins also moonlight. In *E. coli* their moonlighting balances the rate of their own synthesis with the rate of transcription of ribosomal RNAs. When ribosomal RNA is abundant, the proteins bind to it. But when the ribosomal RNA is in relatively short supply, they bind instead to their own mRNAs, thus acting as operon-specific translational repressors. Some are involved in negative regulation. For example, ribosomal protein S2 serves as a negative regulator of the *in vivo* expression of both itself and a translation factor. The list goes on. Nearly 2/3 of the ribosomal proteins from the large (50S) ribosome subunit of *E. coli* have an RNA chaperone activity that ensures proper assembly of the ribosome. And four of the components of NUS, a transcription termination complex, are ribosomal proteins (S4, L3, L4 and L13).

The list goes on and on...

Enough of bacterial fun. The examples of extra-ribosomal employment of ribosomal proteins extend to the eukaryotes, as well. The review we mentioned makes the case that ribosomal proteins also serve as sentinels for the self-evaluation of cellular health. Perturbation of ribosome synthesis frees ribosomal proteins to interface with the p53 system, leading to cell-cycle arrest or to apoptosis.

The list of instances is long, but it could be longer. The authors (http://bit.ly/1RACD91) posit: "*Is this due to a lack of imaginative evolution by cells and viruses, or to a lack of imaginative experiments by molecular biologists?*"

Multifunctionality in proteins ought to give genomic annotators conniptions. Enough to drive them up a multifunctional wall.

Reference

Warner JR, McIntosh KB. 2009. How common are extraribosomal functions of ribosomal proteins? *Molecular cell,* **34**(1):3-11. PMID: 19362532

June 8, 2009

bit.ly/1Llgs5f

54

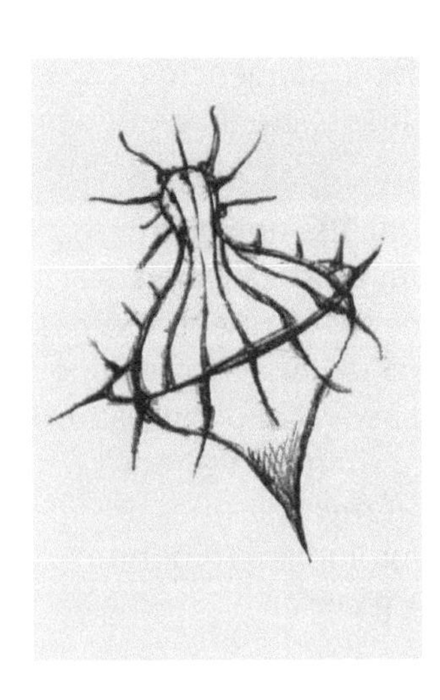

Constructing a Synthetic Mycoplasma

by Shmuel Razin

I would like to raise an exciting issue that has become very hot recently and has been of great interest to me for a long time: the possibility of synthesizing a living cell from its components. My story starts in 1964. I spent then my first sabbatical with Prof. Harold Morowitz at Yale University. Let me first remind you that these were the early sixties, times of revolutionary ideas. It was the time of planning the trip to the moon. We had entertained the idea, promoted by Morowitz, to synthesize a living cell from its components and turn this project into an American National Goal instead of going to the moon. We started by reconstituting the mycoplasma membrane from its solubilized components. Despite early promising results and a rather provocative paper (1.usa.gov/1MhD04c) we published in the *Proceedings of the National Academy of Sciences* on mycoplasma membrane reconstitution, we failed to progress in this direction. Clearly, we were much ahead of the times.

Nevertheless, our idea that was considered crazy 45 years ago has now been revived and appears more plausible than ever. Again, Harold Morowitz was the person who gave it a push at the 5th IOM (International Organization for Mycoplasmology) Congress in Jerusalem in 1984. Morowitz pointed out that mycoplasmas are the smallest organisms capable of independent growth and replication, carrying the smallest genomes of self-replicating organisms. Elucidation of their genes may provide the tool to determine the minimum set of genes essential for life. This talk had a considerable impact and pushed forward mycoplasma genome sequencing. In fact, mycoplasma genomes were among the first bacterial genomes to be sequenced about 10 years later

It appears that *Mycoplasma genitalium* is, thus far, the organism closest to the theoretical minimal cell capable of self-replication. It has been of great interest to define the minimal set of essential genes in *M. genitalium* by selective inactivation or deletion of genes, testing the effects of each of these manipulations on survival and replication under defined conditions. To make a long story short, Craig Venter and his associates have established the fact that mycoplasmas are devoid of cell walls, which facilitates the transfer into the minimum number essential for life for *M. genitalium*: 381 genes.

Venter's early idea was to insert this minimal set of genes into ghosts of *M. genitalium* cells depleted of their original genomic DNA. This would result in a new, still hypothetical synthetic organism, already named by Venter *Mycoplasma laboratorium*. The very ambitious project of creating a synthetic mycoplasma has already resulted in heated discussions and controversy, published mostly in *Science* but getting also into the daily press. Some scientists are being alarmed by the idea that Venter plays God, creating a new organism. But you may understand my feeling of satisfaction at the great interest this creates in mycoplasmas, my baby organisms. Thus, I tended to favor Craig Venter's initiatives.

Venter does not really need my support and with his exceptional energy and money, his Institute (The J.Craig Venter Institute, Rockville, Maryland) did already succeed in assembling a complete synthetic *M.genitalium* genome. This project presents an impressive technical feat, joining together chemical synthesis and recombination cloning methodology. It certainly can be considered as an important milestone in synthetic biology.

Venter's group has progressed also in a parallel direction by showing that a wild-type mycoplasma genome in the form of naked DNA can be propagated by polyethylene glycol-mediated transfer into recipient mycoplasma cells displacing the recipient cell's original genome. This exploits cell of external naked DNA. Thus, what remains to be done is to use this technique to introduce the synthetic *M. genitalium* genome into mycoplasma cells and to search for cells in which the synthetic genome replaced the original genome.

In summation, this somewhat futuristic biotechnological project will be a powerful new approach to understanding biology. In terms of practical applications, one can envision a future in which the complete redesign and prototyping of genomes for industrial microbes leads to new, better and cheaper bio-products.

Shmuel Razin is Jacob Epstein Professor of Bacteriology, Emeritus, at the Hebrew University-Hadassah Medical School.

May 4, 2009

bit.ly/1KluEHy

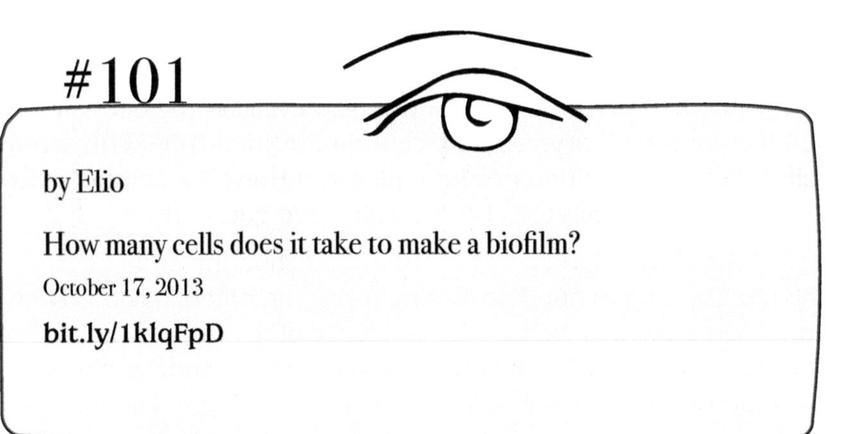

#101

by Elio

How many cells does it take to make a biofilm?

October 17, 2013

bit.ly/1klqFpD

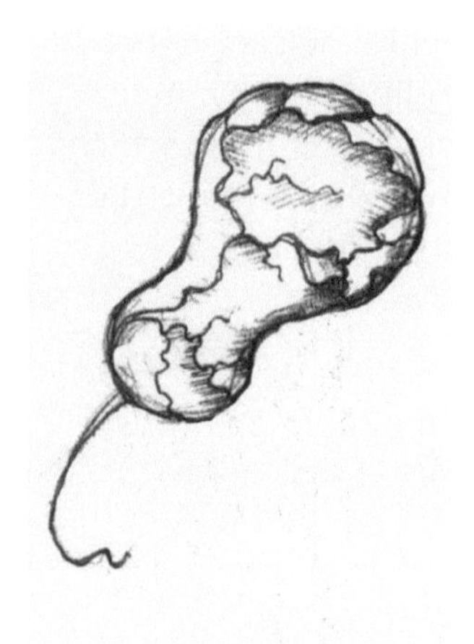

55

By Chance and Necessity: The Role of the Cytoskeleton in the Genesis of Eukaryotes

by Daniel P. Haeusser

How Did Eukaryotes Evolve from Prokaryotes?

One of the most exciting and enduring obscurities of biology lies in the early stages of the evolution of "our" eukaryotic cells. The endosymbiotic theory accounts well for the present existence of the mitochondrial and chloroplast organelles of eukaryotes. Although there is evidence for present-day inter-bacterial endosymbiosis, the details of the route leading to the establishment of organelles remain enigmatic.

Even murkier is the question regarding the origin of the nucleus. While prokaryotic cells with organelles arguably exist and scientists have identified *Planctomycetes* that enclose their DNA with an internal membrane continuous with the cell membrane, a truly independent membrane-bound nucleus prevails as the defining hallmark of the eukaryotes. (A recent report [bit.ly/1O6E8hE] even calls this textbook difference into question! However, another study [bit.ly/1PQ02G3] calls these conclusions into question.)

Unlike the clear bacterial origin of the mitochondria and chloroplast, it turns out that no single existing model has received broad acceptance to explain the existence of the nucleus. Scientists have speculated that it could have arisen from an endosymbiotic event between an archaeon and a bacterium (though who engulfed whom is also uncertain). Some have proposed that an autonomous internal gathering of cell membrane as in *Planctomycetes* formed the eventual independent nuclear organelle, and yet others have even suggested infection by an enveloped virus as playing a role.

Not only are details on the evolution of these eukaryotic innovations (nucleus and mitochondria or chloroplast) shrouded in mystery, but just as unclear is the temporal relation between their developments. Which came first, the nucleus or the mitochondrion?

A recent online article (bit.ly/1k8J20x) by Ed Yong for *Nautilus Magazine* gives a comprehensive summary of one of the prevailing hypotheses proposed to answer that question. The general crux is that the relatively sudden marriage of two prokaryotes into one stable individual cell occurred just once in the evolutionary

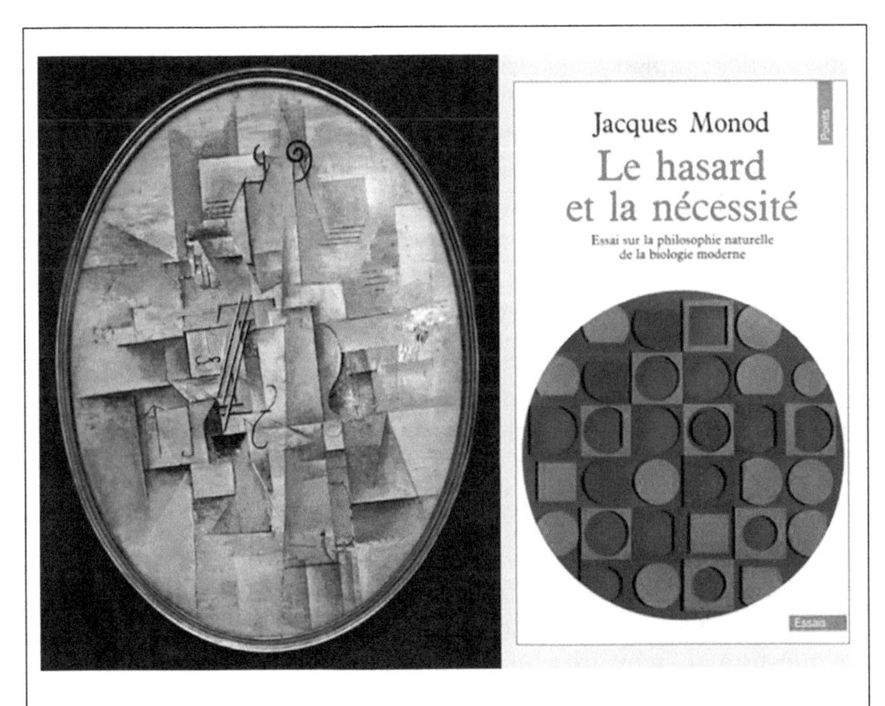

(Left) Pablo Picasso's Violon (Violin) 1911-12, Kröller-Müller Museum, Netherlands. Just as the audience's image of a violin is destroyed through deconstruction within the oval frame to achieve Picasso's art, so too may a bacterial cytoskeleton have been disrupted under the influence of archaeal-evolved factors following endosymbiosis. (Right) Chance and Necessity by Jacques Monod. Borrowing Monod's title, the chance merging of two distinct systems for cytoskeletal regulation may, by necessity, have influenced eukaryogenesis.

Sources: 1: Wikipedia, 2: Goodreads

history of life on Earth, and that allowed the emergence of an entire new branch on the tree of life. Here, the engulfment of a bacterium by an archaeon permitted the development of mitochondria. This union augmented and expanded the metabolic and energy generation capabilities of the new organism, thereby permitting genome size expansion and organelle proliferation (compartmentalization). All this gave sudden rise to the establishment of the hugely differentiated eukaryotic lineage.

Some pieces of evidence detailed in the *Nautilus* article are consistent with a model where mitochondrial acquisition "licensed" the genesis of eukaryotes. Yet, the jarring thing about this mitochondria-centric hypothesis is that it says nothing about the nucleus. The big problem with mitochondrial acquisition alone as the driving force of eukaryogenesis is that such a development is not a simple matter of metabolic and energy concerns. As recognized in Yong's article, prokaryotes also largely lack eukaryotic "architecture," leaving them "*forever constrained in size and complexity*," an idea also discussed in an essay (bit.ly/1PTz67F) by Kevin Young. The mitochondrion alone is not a ready system for getting things around a cell as it increases in size and complexity. The cell would still lack the architecture and machinery to be released from the limits of simple diffusion, a significant barrier between the small prokaryotes and the (usually) larger eukaryotic cells.

In short, the mitochondria-centric hypothesis for eukaryotic genesis ignores the important uniqueness of the eukaryotic cytoskeleton. So, let's take the cytoskeleton into consideration.

Why Are Prokaryotes Different from Eukaryotes?

A superb, recent essay (bit.ly/1P42Uxo) by Julie Theriot addresses these issues with a focus on comparing eukaryotes to bacteria (not to archaea). Theriot's essay probes deeply and meaningfully into the question of "why are prokaryotes and eukaryotes different." With expertise in the cytoskeleton, her ruminations on this "why" question rest on that dynamic architecture of eukaryotes that permits them to reach proportions of growth and organization rarely seen in prokaryotes.

To summarize her major points: While we now know that the cytoskeleton is present and broadly conserved across all domains of life (with diverse proteins in the actin, tubulin, and intermediate filament families), bacteria notably lack a special class of cytoskeleton-related proteins: *the nucleated filament assembly factors* and *the motor proteins.* Eukaryotes make ample use of these tools. By providing a mechanism for polarized assembly that permits motor protein directionality and inherently oriented filament growth, the eukaryotic cytoskeleton is the architectural machinery that achieves functions such as chromosome movement, vesicle and organelle transport, and endocytosis. These, in turn, help the cell attain sizes and complexities that otherwise would be unattainable.

Theriot correctly points out that, in theory, bacteria are capable of making such cytoskeleton-related factors. Indeed, several pathogens contain genes for proteins that alter and control cytoskeleton assembly including through nucleation. But bacteria export these proteins for use on the host never (that we know of) for their own cytoskeleton. So why didn't they learn to use their cytoskeletons in the same way to gain some of the benefits that the eukaryotes enjoy? Why not follow the eukaryotic route?

Possible answers to these questions receive due consideration in Theriot's essay, but what I found most fascinating is her observation that bacteria have a trove of genes that encode different varieties of the major cytoskeleton protein families, particularly with actin homologs. It seems as though bacteria have evolved one distinct homolog for each cellular function. Eukaryotes, by contrast, mostly contain just one type of each family member, which is then specifically regulated to perform several distinct functions. Could this explain why bacteria don't use their cytoskeleton like eukaryotes?

With Theriot's essay and the *Nautilus* article in mind, I wondered whether the existing unique properties of the eukaryotic cytoskeleton could fit into models for those cryptic early steps in the evolution of eukaryotes from prokaryotes. So, let's circle back to the question of eukaryogenesis, but now with the cytoskeleton in mind.

"Every Act of Creation Is First an Act of Destruction" – Pablo Picasso

What if the first endosymbiotic merging that permitted the outburst of the eukaryotic lineage were not that of mitochondrial acquisition (a bacterium engulfed by an archaeon), but rather the reverse, something more akin to what has been considered a possible origin of the nucleus, the engulfment of an archaeon by a bacterium.

Theriot's essay avoids discussion of archaea, perhaps because we know precious little about them. In visual appearance, archaea and bacteria are look-alikes, but archaea contain genetic and mechanistic facets that are both bacteria-like and eukaryote-like. The cytoskeletons of archaea are similarly mixed: while many encode bacterial components such as *ftsZ*, their overall cytoskeleton composition is closer to the eukaryotes'. Crenactin, for instance, is phylogenetically nearest to canonical actin rather than to the myriad actin homologs found in bacteria. We don't know enough about archaeal cytoskeleton control yet to figure out if they have nucleated assembly factors, motor proteins, or any of the other unique properties of the eukaryotic cytoskeleton. But given the presence of other rudimentary eukaryotic systems in archaea, such as ubiquitination, it's reasonable to hypothesize that some archaea may employ and control their cytoskeletal architecture in ways closer to those of eukaryotes.

So then, imagine if a bacterium with its cytoskeleton composed of multiple, monofunctional actin/tubulins were to engulf an archaeon with a relatively simpler cytoskeleton consisting of a single member of the actin and tubulin families each. Were the engulfed archaeon capable of growth and further replication within its new host, some cells would die with time, leaking archaeal contents into the bacterial host cytoplasm. I propose that this could include "eukaryotic-like" cytoskeletal proteins and regulatory system components like proto-motor proteins and nucleation factors.

The presence in the same bacterium of multiple actin and tubulin homologs carries the risk of cross talk between the regulator and those homologs. If so, this may preclude or at least discourage the evolution of task-specific regulators that affect assembly like nucleation proteins, or factors like motor proteins. Thus, if the cell needs to nucleate one actin homolog involved in plasmid segregation, it may not want to have to nucleate the other actin homolog, e.g., one involved in maintaining cell shape. Unique factors would then need to evolve for each cytoskeleton homolog, a demanding task. On the other hand, the "single member" system of the archaea may allow for the innovation of these kinds of regulatory systems. They could comprise a small set of regulators whose activity can be modulated depending on where and when they're needed. Inside their bacterial host, these archaeal regulatory factors are now free to act via crosstalk on the bacterial homologs.

This potential for disastrous cross-talk of a single regulator on different homologous cytoskeleton family members in a single cell would also explain why several bacterial species appear perfectly capable of evolving eukaryotic-like cytoskeleton control for export during pathogenesis (in *trans*), yet eschew adoption of such systems in *cis*. Their relatively broad repertoire of highly similar cytoskeletal components simply makes it too risky a proposition.

The Picasso quote that titles this section could be equally apt for this proposed version of eukaryogenesis To summarize and repeat: The chance mixing of two distinct modes of cytoskeleton regulation in merged organisms (one with many similar members with unique functions and the other with single members with broad functions) would potentially result in a war for control. Because of regulatory crosstalk, one system would need to eventually win out. Perhaps the products of the endobacterial archaeal genome interfered with, and ultimately destroyed, the bacterial cytoskeletal system, creating a novel one in its place.

An archaeal cytoskeleton with rudiments of nucleation and motor proteins, but now within a bacterium with its own unique genes for metabolism, creates a novel organism with the architectural system needed for further evolution. This could include the compartmentalization of a newly merged genome under the control of an archaeal-derived cytoskeleton. With the acquisition of energy-generating organelles and further compartmentalization, larger physical dimensions and a bigger genome size could be attained. More stable, these early eukaryotes would displace any amitochondrial species and, with a finer tuning of their cytoskeleton, develop more innovations.

According to this model I'm proposing, the key event in the eukaryogenesis is, to reference Monod (bit.ly/1Wh3XzR), the *chance* merging of two systems by the *necessity* of regulatory cross-talk gaining dominant control over homologs of the other system. Above all, such a model predicts that we should be able to find existing archaea with genes that encode nucleation factors and/or motor proteins used to control the functions of its own cytoskeleton. To my knowledge they have not yet been found, but I don't think that we've looked particularly hard either.

Are there problems with this line of thinking? Are there other considerations? Most importantly, what kind of experiments do you think would currently be possible to address these speculative models? The mystery of eukaryogenesis still has a lot of questions, and like with many things in science, I don't think that our curiosity will soon be satisfied.

Daniel P. Haeusser is an Assistant Professor at Canisius College and an Associate Blogger at STC.

April 14, 2014

bit.ly/1KluGPz

#40

by Richard Moxon

Can you think of any human disease that does not have a possible microbial component?

November 6, 2008

bit.ly/1M3pAz3

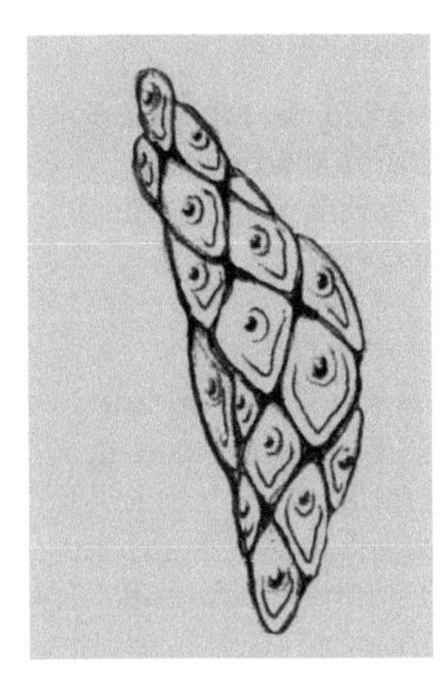

56 The Gram Stain: Its Persistence and Its Quirks

by Elio

What is more emblematic of our science than the Gram stain? Since its invention 130 years ago, it has been in frequent and continuous use. It conveniently places most bacteria into one of two groups, the Gram-positives or the Gram-negatives. Gram staining is cheap, effective, quick, and relatively easy to interpret. Its most useful application is in the clinical setting. When examining a smear of, say, pus from an abscess, this stain often allows to include for consideration roughly half the clinically relevant bacterial species while excluding the others. Or consider a patient with meningitis. Here, speed is of the essence because treatment must be initiated right away. A Gram stain of the spinal fluid may reveal within minutes the presence of Gram-positive cocci, (probably pneumococci), Gram-negative cocci (almost certainly meningococci), or Gram-negative slender rods (most likely *Haemophilus influenzae*). This can make a critical difference in the choice of antibiotics that have to be administered in great haste. However, when it comes to characterizing the bacteria in an environment, its usefulness diminishes, in part because it is not always in step with taxonomy, which I'll discuss below. And there yet is another side to this story. Quite a few bacterial species that stain positive early in the growth of a culture become Gram-negative later on. Does this detract from the value of Gram's method? It may, but not in the hands of a person experienced in its use.

A Bit of History: How to Stain Bacteria in Tissue Slices

In the early days of bacteriology, staining was a big deal (culturing was just coming into its own). Staining procedures were aided by the availability of strong dyes, the synthetic aniline dyes, that were first made around the middle of the nineteenth century. Staining bacteria on a smear with these dyes worked fine, but visualizing them in tissue slices—which is one way to fulfill Koch's first postulate—was another matter. Here, all sorts of cellular and extracellular material also take up the dyes used, which makes it nearly impossible to distinguish the small bacteria. The problem, it was realized, could be solved by using ways to selectively stain the bacteria. One way to do it could be to decolorize the other stuff. The first successful technique based on this principle used acid to differentially remove the stain. Fortuitously, this method was developed to distinguish tubercle bacilli, which uniquely retain a dye after acid treatment (hence, they are "acid fast"). Other bacteria are bleached by acid along with what's in the rest of the sample.

Finding ways to differentially stain other bacteria in tissues required a lot of work and all sorts of trials. One strategy that clearly came to mind was to use mordants, chemicals that help "set" a dye. This was standard practice in the textile industry, the principal user of these synthetic dyes and the developer of mordants for intensifying the colors. Many possible mordants were used to dye bacteria as well, e.g., phenol in the acid-fast staining procedure. These methods were known to the Danish physician H. C. Gram (1853–1938) who wanted to see the bacteria in slices of lung tissue from patients who had died of pneumonia. He did not have in mind distinguishing one kind of bacterium from another, the main purpose today for the procedure that bears his name.

Gram tried as a mordant a common iodine solution, Lugol's, which he almost surely had within arm's reach. When using a violet stain, he soon found out that adding the iodine solution precipitated the dye. In other words, it made a mess. Hoping to clean up the preparation, Gram tried flooding it with ethanol. He found that, if he did it carefully, the tissue components were decolorized, whereas some bacteria retained the violet dye and others did not. The colorless ones were now hard to see, but adding a "counterstain" of a different color (red) made them readily visible. So, Gram-positives become "purple," Gram-negatives "red." Being a modest man, Gram stated: "*I have therefore published the method, although I am aware that as yet it is very defective and imperfect; but it is hoped that also in the hands of other investigators it will turn out to be useful.*" How true is the last statement! It bears stressing that Gram's discovery was anything but accidental, despite what has been stated in some textbooks.

How Does It Work?

The crux of the Gram stain is the exposure to alcohol (or to the other organic solvents that are often used) during decolorization. For the Gram-negatives, most of which possess a thin peptidoglycan cell wall, this damages the cytoplasmic membrane and allows the escape of the dye-iodine complex. The Gram-positives, on the other hand, typically have a thick peptidoglycan layer that somehow impedes the passage of alcohol or other solvents. It is thought that perhaps solvents render the envelope impermeable by dehydrating the peptidoglycan—a good thought that appears in some writings but for which I have not found the primary sources. I would appreciate it if someone could point me to them.

One further thought. Almost universally, the Gram stain is carried out on heat-fixed smears on a microscope slide. This time-tried step is of course simple and convenient. However, heat and drying may further weaken the cell envelope components, thus adding to the loss of Gram-positivity. For the sake of gaining further understanding of the mechanism of the Gram stain, perhaps studies should be carried out on unfixed cells. I expect that under those conditions some cells may not take up the dye or the iodine solution, but this could probably be tampered with. Anybody know of such studies?

The Gram Stain Encounters Taxonomy

In a simpler world, all bacteria with a single membrane, the *monoderms*, would be Gram-positive, all those with two (the *diderms*), Gram-negative. Put taxonomically, all Firmicutes (e.g., *Bacillus, Clostridia, Staphylococcus, Streptococcus*) and Actinobacteria (e.g., *Corynebacterium, Mycobacterium, Streptomyces*) would be Gram-positive, all the rest (e.g., *Escherichia, Haemophilus, Pseudomonas, Neisseria*), Gram-negative. Hah! Were it that simple! Each of these groups includes outliers that stain opposite from the expected way. An illuminating essay on evolutionary considerations that helps understand this state of affairs was recently written by Jeff Errington.

Most monoderms fall conveniently in the Gram-positive camp all right, but there are important exceptions. First, the Firmicutes (so called for the "firmness" of their thick peptidoglycan cell wall) encompass the wall-less mycoplasmas and their allies, bacteria that cannot be stained by the Gram stain at all, which makes the point moot here. Next, some typical Firmicutes (the Negativicutes) stain Gram-negative. To make things worse, at least one firmicute, Acetonema, appears to have the prototypical Gram-negative property, an outer membrane (a fascinating story for some other time, but the reason I got into this subject in the first place). Having a thin peptidoglycan layer may place whole phyla (e.g., the green non-sulfur bacteria or Chloroflexi) within the Gram-negatives, despite being monoderms.

When it comes to consistency, the diderms fare no better. Some, such as Deinococcus, stain Gram-positive because they have an unusually thick peptidoglycan layer, despite being diderms and possessing the two membranes typical of Gram-negatives. And if this were not bad enough, lesions that affect cell integrity can render otherwise Gram-positives to stain as Gram-negatives. Authentic Gram-positives such as members of the genus *Bacillus* will gradually lose their Gram-positivity and some of their peptidoglycan cell wall after they stop growing. Note that I have not mentioned the Archaea, which are mainly Gram-negative, with some interesting exceptions.

Gram Variability

Clinical microbiologists are used to living with "Gram indeterminate" bacteria. The matter of Gram variability is actually quite a bit more intricate than I made it sound. The late Terry Beveridge, whose stellar contributions to microbial cell structure were celebrated in these pages (bit.ly/1O6F05K), studied this phenomenon in detail, using a variety of organisms. He cleverly synthesized an iodine analog, trichloro(eta 2-ethylene)-platinum(II), that is electron scattering, hence could be used to look at sections of Gram-stained bacteria in the electron microscope. Read his paper (1.usa.gov/1GH-V2QR) for details. In brief, anything that damages or weakens the cell envelope may turn Gram-positives negative. This includes the thinning of the cell wall peptidoglycan layer that happens as some cultures age, or the weakening of the protein S-layer, with which some bacteria and many archaea are endowed. Gram staining imposes great stress on the integrity of such organisms, so it's not surprising that Gram-positivity can be readily lost. Note that it's relatively easy for a Gram-positive to become Gram-negative, whereas the converse is not the case. This may not be the whole story because fragments of Gram-positives have been reported to stain positive, possibly suggesting an alternative mechanism for the Gram stain.

Summary

Let me wrap this up by saying again that most monoderms stain Gram-positive, most diderms, Gram-negative. However, there are important exceptions. Knowing that an organism is a mono- or a diderm doesn't tell you for sure what its Gram staining properties will be. Some monoderms stain negative, some diderms, positive. What seems to count is how thick and intact the peptidoglycan layer is. And, I should add, the Gram stain is not useful to distinguish the Archaea, which don't have peptidoglycan and stain in a seemingly unpredictable manner. It's a miracle that despite these shenanigans, the Gram stain remains a major tool in the clinical microbiology lab. And, for all the exceptions, it proved to be a key tool in the elucidation of central properties of the bacterial cell.

I thank Jeff Karr, the ASM archivist, for supplying original material for this paper.

February 11, 2013

bit.ly/1NliCCz

57

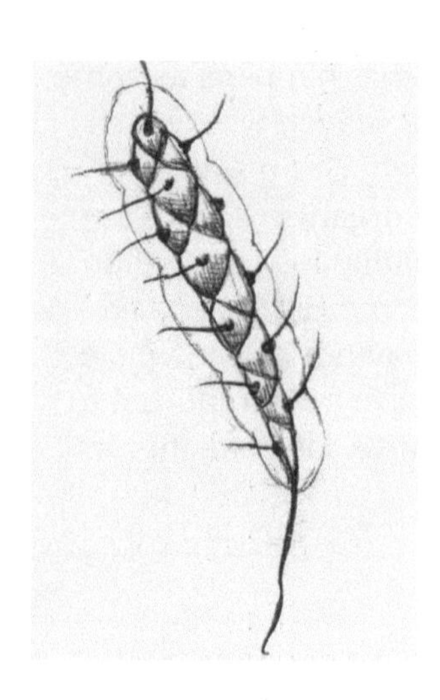

On the Continuity of Biological Membranes

by Franklin M. Harold

Thirty years ago, Günter Blobel of the Rockefeller University published a short paper entitled "Intracellular Protein Topogenesis," which laid the conceptual foundations for our understanding of how cells build membranes. To serve their functions, peripheral and integral proteins must be inserted into the right membrane with the correct orientation, and most of the article focused on the manner in which this may be achieved. But it also underscored two startling implications of the proposed procedure: first, that every membrane must be derived from a pre-existing membrane; and second, that all extant biological membranes are descendants of the plasma membrane of the first primordial cell.

Blobel's article became a classic, and spawned a small industry concerned with the molecular mechanisms that target proteins to the recipient membrane and then either translocate or insert them. In a nutshell, the information that specifies a nascent protein's disposition is contained in its sequence. One segment of that sequence recognizes a receptor protein embedded in the target membrane, commonly part of the translocon; other segments specify whether the amino acid chain is to be taken clear across the membrane or inserted, and with what orientation. Membrane proteins may be processed concurrently with their translation, or after their production is complete. In prokaryotic cells the proteins are produced and handled directly; in eukaryotic cells they are first inserted into the membrane of the endoplasmic reticulum, and then transferred to their target membrane by cargo vesicles. The details can be found in textbooks of molecular cell biology. What concerns us here is the inference that membrane heredity is a fundamental principle of biology. A functional membrane, studded with a particular set of enzymes, transport carriers, and receptors, can never be generated *de novo*; it must arise from a pre-existing membrane, either by modification (for example, the membranes that surround bacterial spores) or else by growth and division or vesiculation. Moreover, since proteins will only be inserted after interaction with a complementary receptor (and that includes the receptor protein itself), a growing "genetic" membrane propagates its own kind.

The idea that membranes are inherited was by no means novel in 1980; cytologists had been musing on it for two decades. But it was quite another matter to assert that it must be so, that "*omnis membrana e membrana*."

Biology is notoriously so riddled with exceptions that such a sweeping generalization is bound to raise eyebrows: never? Indeed, possible exceptions do crop up from time to time. If this prospect piques your curiosity, take a look at the work of G.H. Kim and his colleagues (here [1.usa.gov/1PTA83g] and here [bit.ly/1GwFeRp]), which describes the astonishing capacity of naked blobs of algal cytoplasm to reconstitute a membrane and resume growth. Blobs endowed with nuclei and a sample of organelles survive transiently in the absence of a plasma membrane (sic!), construct a temporary one made of polysaccharides, and finally produce a proper membrane made of lipids; how they do this is quite unknown and would provide a nice test of Blobel's dictum. In years of reading I have never come across an authentic example of a membrane made afresh, and a query to readers of this blog elicited no response. Like the second law of thermodynamics, the verity that membranes must be grown rather than made rests not on proof positive, but on the absence of any known exceptions.

Even though membrane heredity enjoys general acceptance, it seldom comes up in the literature. The reason, I believe, is that it holds the answer (more correctly, part of the answer) to a question that few scientists are asking, but an important question all the same. As cells grow and divide, the form and arrangement of their internal organelles (many of them membrane-bound, especially in eukaryotes) is quite faithfully transmitted to the next generation; just how does that come about? Time was when transmission of the cognate genes was deemed to be a sufficient reason; though as far back as the sixties scholars such as Boris Ephrussi and Tracy Sonneborn insisted that the inheritance of genes cannot by itself account for the persistence of structural organization. The principle that membranes must be inherited unambiguously sides with those "cytoplasmic heretics" and their followers. Thomas Cavalier-Smith, one of the few prominent scientists to fully embrace Blobel's thesis, puts it clearly and forcefully:

> "*Two universal constituents of cells never form* de novo: *chromosomes and membranes...Just as DNA replication requires information from a pre-existing DNA template, membrane growth requires information from pre-existing membranes—their polarity and topological location relative to other membranes... Genetic membranes are as much a part of an organism's germ line as DNA genomes; they could not be replaced if accidentally lost, even if all the genes remained.*"

Structural order is transmitted jointly by copies of the genes and by architectural continuity. One of the reasons that every cell comes from a preexisting cell is that there is no other way to make a membrane.

Not only are membranes passed from one generation to the next, they are remarkably persistent on the evolutionary timescale. This is most vividly illustrated by the membranes of mitochondria and chloroplasts, both of which descend from endosymbiotic eubacteria. No one knows for sure how ancient these partnerships are, but since all extant eukaryotes apparently derive from a common ancestor endowed with mitochondria, this one must go back one or two billion years, and possibly more. Chloroplasts were probably acquired later, but even that event dates back to at least 600 million years ago, and probably longer. In the course of their "enslavement" and reduction to the status of organelles, most of the endosymbionts' genes were either transferred to the host's nucleus or lost altogether. Nevertheless, the membranes of both organelles clearly proclaim their bacterial ancestry, both in their chemical composition and in their morphology. In the case of chloroplasts, the number of membranes that surround the organelle tracks the history of successive episodes of symbiosis. The chloroplasts of green plants and algae, red algae, and glaucophytes, offspring of the primary endosymbiosis, are encased within two bilayer membranes, derived respectively from the inner and outer membrane of the cyanobacterial endosymbiont. But the chloroplasts of many other photosynthetic protists are enveloped in three or even four bilayer membranes, which are believed to report a history of secondary or tertiary endosymbiosis: cases in which a non-photosynthetic protist engulfed and assimilated a photosynthetic algal cell in its entirety. (For a review of this complicated story, go to 1.usa.gov/209rjXI. Membranes are not immune to evolutionary change; they are subject to radical alteration and reduction of function, and may also be lost altogether. A striking example of membrane transformation is supplied by hydrogenosomes, metabolic organelles of anaerobic protists, which are thought to derive from mitochondria with the loss of the respiratory chain; even more extreme reduction produces residual membranous bodies known as mitosomes. It seems to be the membrane-bound compartment, not its functional proteins, that has the propensity to endure; sometimes what matters is the bag, not its contents.

Organelles make an impressive example of the persistence of membranes, but one could wish for more of them. A likely one comes from the Archaea, whose membranes all display a distinctive complement of lipids and ion-translocating ATPases, even though their environments range from volcanic hot springs to the open ocean and the stomach of cows; it cannot be natural selection alone that maintained the archaeal signature! As for eukaryotic cell membranes, they are evidently of dual origin. Those of mitochondria and chloroplasts were inherited from the endosymbionts; the provenance of the others is in dispute, but the most plausible hypothesis at present is that the membranes of the nucleus and endoplasmic reticulum represent infoldings of the host's plasma membrane. Let me reserve this minefield for a future comment; in the meantime, should you know of other examples of membrane heredity, do please let me know!

The doctrine that it takes a membrane to make a membrane has profound implications for the origin and evolution of cells. First, if the molecular machinery of protein translocation is required to put in place integral membrane proteins, how could functional membranes have existed before there were translocons, let alone proteins? If every membrane must grow from a pre-existing membrane and reproduces (or modifies) its topology, how could this lineage have begun when there were no membranes to copy? This is one of the many chicken-versus- egg paradoxes that bedevil the mystery of cellular origins, and one that is not at all laid to rest by postulating that an RNA World preceded the DNA/RNA/Protein World that we inhabit today. Second, persistence of membranes carries a strong hint that the conventional view, which derives the first cells from aggregation of biological molecules produced by abiotic chemistry, is fundamentally mistaken. Instead, life must have been in some degree cellular from the very beginning: the product of co-evolution of genes, catalysts and membranes in a structured setting, as Cavalier-Smith has argued for many years.

The genesis of membranes, and of cells, quite passes understanding; nevertheless, the field presently displays a ferment of experiments and ideas that must owe something to the relentless challenge from advocates of intelligent design. In his original paper, Blobel suggested that the first precursors of cellular life were lipid vesicles that had formed spontaneously in the primordial broth. Their outer surfaces provided capturing devices for the coalescence of ancestral molecules involved in replication, transcription, and translation, as well as metabolic enzymes, all assumed to be present in the surrounding medium. Translocation of molecules or segments thereof across the lipid bilayer into the interior phase would have evolved at this stage. Note that the polarity of these protocells would have been inside-out relative to cells as we know them (enzymes and ribosomes on the outer surface, not in the lumen). So Blobel sketched a scheme to make them invaginate, close up into a "gastruloid" enveloped by a double membrane, and thus assume the familiar polarity of contemporary cells, with all the machinery on the inside. These ideas have been adopted by Cavalier-Smith (1.usa.gov/1MSHejs), who explains in much detail how "obcells" would have formed, functioned and at last turned outside-in. An important element of Cavalier-Smith's thinking is that the first true cells were enveloped in two bilayer membranes; cell evolution must, therefore, have begun with "negibacteria," i.e., Gram-negatives. It all makes sense, if you can believe that at least the rudiments of life's molecular machinery took form out there in the soup, with inorganic polyphosphate tossed in as an energy source. But the obcell hypothesis has never caught on, presumably because readers judge it to be just too implausible, and so do I. A recent re-formulation (http://1.usa.gov/1k8LiEZ) by Griffiths addresses some of the difficulties, but falls short of eliciting a "Eureka!" reaction.

But what are the alternatives? Lipid membranes can form abiotically (ingredients are even found in carbonaceous meteorites), and they can encapsulate macromolecules such as RNA and a polymerase. But the spate of recent publications in this vein (here [bit.ly/1kPTlao] and here [1.usa.gov/1PQ0Jzm]) never touches on the issues raised by membrane protein topology, nor on membrane inheritance. So let me instead draw attention to a very different idea that has languished on the fringes of serious science ever since the geochemist Michael Russell first articulated it two decades ago, but is now gaining traction. Believers find the cradle of life (here [bit.ly/1k8LDHE] and here [1.usa.gov/1KEzQGK]) in the nooks and crannies of porous mineral deposits formed at the edges of submarine hydrothermal vents, specifically warm and alkaline ones such as the Lost City Field. Alkaline hydrothermal vents make an attractive venue for the early stages of chemical evolution: sequestered spaces, reasonable conditions, and an ample supply of precursor molecules including hydrogen gas, methane, and small organic compounds. Geochemistry even supplies a potential energy source: the large difference in pH between the alkaline vent fluids and the acidic bulk water (as much as four units). It does not strain credulity to suggest that among the products of vent chemistry may have been amphipathic molecules that aggregated upon surfaces and occasionally generated primitive membranes. If (and what a big If that is!) chemical complexity burgeoned in the honeycomb to the point of simple metabolism and heredity, some of those membranes may have grown, propagated their kind and come to enclose cell-like bubbles with the correct polarity. In the fullness of time, could some of those bubbles have escaped from their inorganic hatchery, setting forth to seek their fortune and inherit the earth? Might the fundamental differences in lipid chemistry between Eubacteria and Archaea report separate origins from different hydrothermal mounds? Well, let's not get carried away. The notion that cells were born of hydrothermal vents also has multiple pitfalls, notably the lack of any obvious driving force to channel chemical evolution in the direction of biological functions;

but it is a fantasy well worth pondering.

This is all good, clean fun—as long as we prize the doubt, keep a sense of humor, and do not pretend to the authority that comes only with hard, experimental science. Karl Popper taught us that science advances best by the interplay of conjecture and refutation; unfortunately, students of cell evolution do the former rather better than the latter. Even in this Age of Omics, when it comes to making sense of the incomprehensible we can only place our trust in tales of the imagination.

Frank Harold is an affiliate professor in the Department of Microbiology, University of Washington Health Sciences Center. Now retired, he remains engaged with science as a writer and unlicensed philosopher.

March 01, 2010

bit.ly/1QNSocE

#43

by Mark Martin

How might the botulism bacillus, *Clostridium botulinum,* benefit from making botulinum toxin? (This question can be extended to the tetanus bacillus and its toxin, as well as to others.)

January 15, 2009

bit.ly/1W2T18T

58

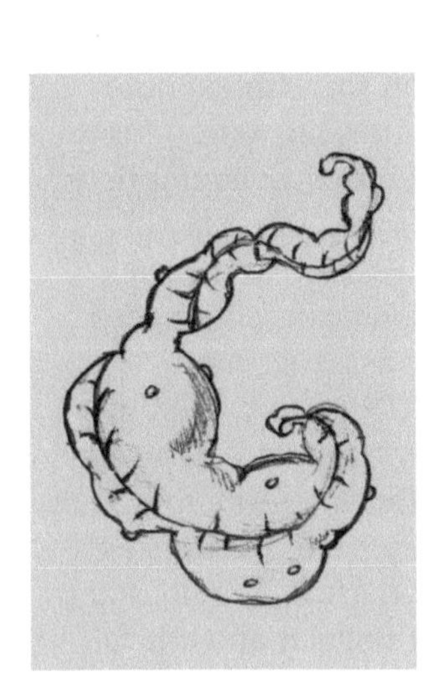

Energetics of the Eukaryotic Edge

by Franklin M. Harold

The most conspicuous feature in the landscape of cell evolution is the tremendous rift that separates eukaryotes from prokaryotes. This is not apparent at the level of ribosomal RNA sequences, or of molecular biology in general, but leaps to the eye of anyone intrigued by form, function, and evolutionary potential. Prokaryotes have been biochemically most inventive, and found access to all the practicable energy sources our planet has to offer. It's a good rule of thumb that, if a chemical reaction yields sufficient energy to support life, a prokaryote exists that exploits it. But when judged by their morphology and organization, prokaryotes seem to have advanced little beyond their fossil ancestors of 2 to 3 billion years ago. Some, it is true, have attained structural and behavioral complexity beyond the norm: cyanobacteria and planctomycetes with their internal membranes come to mind, and so do myxobacteria with their elaborate fruiting bodies and wolf-pack hunting habit. Still, these pale by comparison with even the plainest of eukaryotic protists, whose cells are typically a thousand times larger and stuffed with functional machinery. It is almost seems as though the prokaryotes made repeated starts up the ladder of complexity, but always fell short. By contrast eukaryotes, despite their meager metabolic repertoire, burst whatever constraints hampered prokaryotes to experiment with the opportunities afforded by greater cell size and more elaborate organization. Just what is it that made the eukaryotic mode of life so much more "evolvable" than the prokaryotic one?

Evolvability is an abstract term that refers to "*the capacity to generate heritable, selectable phenotypic variation*" (1), and sometimes more broadly to the propensity to evolve novel structures (2). Marc Kirschner and John Gerhart, whose definition I have just appropriated, credit the conspicuous evolvability of metazoa to the flexibility of the developmental processes that mediate between genotype and phenotype. Loose reins also rank among the significant differences between prokaryotic cells and eukaryotic ones: as a rule, the latter have many more genes and much extra DNA, more tolerant regulation of gene expression, many more moving parts, and a more fluid physiology. These are surely important differences, but they look like consequences of the distinctive evolutionary strategies pursued by the two kinds of cells, rather than the cause thereof. Nor does it make sense to attribute the difference to some singular and critical invention, such as nuclear membranes or linear chromosomes. Though characteristic of eukaryotic cells, both are found in some prokaryotes as well; even endocytosis has recently been documented in planctomycetes (3).

A more convincing argument links the gulf that divides prokaryotes from eukaryotes to their distinct ways of organizing the production of useful energy. In prokaryotes energy generation is a function of the plasma membrane, whereas eukaryotes assign it to specialized components, the mitochondria and chloroplasts. This greatly expands the membrane surface available for energy transduction by ion currents, while freeing the plasma membrane for other tasks. However, as Nick Lane and William Martin explain in a stimulating paper (4), this is just the beginning of the story, for that anatomical difference has unexpected evolutionary implications. In prokaryotes, the energy generated by each individual cell supports the output of its own genome; as much as three quarters of that energy is required just to express genomic information by way of protein synthesis. By contrast, in eukaryotic cells the energy produced by hundreds or even thousands of mitochondria, each one a prokaryotic power pack, is put at the service of a single central genome. In consequence, a eukaryotic genome governs far more energy than a prokaryotic one. This allowed eukaryotic genomes to expand, laying the foundations for the spectacular difference in cellular complexity. Lane and Martin argue that the rise of the eukaryotes was due to the acquisition of mitochondria at a very early stage in their evolution, and furthermore, that the only way structurally and functionally complex cells could have made an appearance was by enslaving endosymbionts.

Let a few numbers taken from that paper bolster the thesis. When expressed on the basis of mass, the mean metabolic rates of aerobic eubacteria and protozoa are not very different: 0.19 and 0.06 W/g, respectively (one watt corresponds to one joule per second). But their cellular masses are very different indeed, typically 2.6×10^{-12}g for the bacterium and $40{,}000 \times 10^{-12}$ g for the protozoan. In consequence, a protozoan cell has much more power at its command than a bacterial one: 2300 pW per cell, compared to 0.49 pW. Eukaryotic cells made use of that abundant energy to expand their genomes by orders of magnitude: the mean haploid DNA content is 6 megabases for a prokaryotic cell, 3000 megabases for a protozoan. (mean gene numbers are 5000 and 20,000, respectively, though both prokaryotic and eukaryotic genomes vary over a wide range). Even so, the energy available per gene is far greater for eukaryotes than for prokaryotes, 57 fW versus 0.03 fW per gene (other ways of expressing the difference make the disparity even larger). The mitochondria that power a eukaryotic cell still harbor a minimal genome, a remnant of the genome that came in with the ancestral endosymbiont; most of those genes were either lost or transferred to the host's nucleus. Taking organellar genes into account makes little difference to the estimate of the power available per gene. Evidently, eukaryotic cells have "energy to burn," more than sufficient to evolve many new protein families, to explore sophisticated ways to regulate their production and put them to work in elaborate physiological processes that would be beyond the means of prokaryotic cells.

Lane and Martin draw an instructive comparison between the energy budgets of a protozoan cell and a giant prokaryote, such as *Epulopiscium fishelsonii*, several times the size of *Paramecium*. A cell of *Epulopiscium* is a consortium of some 200,000 full-fledged nucleoids, each reliant upon a share of the energy generated by the communal plasma membrane. Reproduction requires the cell to duplicate 760,000 megabases of DNA, compared to 6000 megabases for a protozoan cell of comparable size. *Epulopiscium* obviously generates sufficient energy to reproduce, but will have none to spare on a quest for structural and functional complexity.

Could one imagine some way for prokaryotes to compartmentalize their energy production in small pods, and thereby augment the genome's power supply without going to all the bother of domesticating endosymbionts? Yes, one could, but bacteria have not followed that path, and Lane and Martin argue that in reality no such option was available. The reason is that the small genomes retained by both mitochondria and chloroplasts perform an essential function. Maintaining a steady flux of electrons through the respiratory (and photosynthetic) redox chains requires the cell to make adjustments and repairs to individual organelles as needed, and that can only be done under the control of a genome localized to the particular organelle (5). That kind of architecture can only be achieved by starting with endosymbionts, enslaved and put to work as components of a larger community.

These ideas have implications for one of the thorniest issues in cell evolution, the origin of eukaryotic cells. In the conventional view, mitochondria came late into a phagocytic cell, descended from the eubacterial stem, that was already well on its way to attaining eukaryotic organization (6,7). Lane and Martin reject that position in favor of the hypothesis that the first step in eukaryogenesis was the fusion or merger of an archaebacterium with a eubacterium (8,9). The former made large contributions to both the cytoplasm and the nuclear genome, the latter became the precursor of mitochondria, and their increasingly intimate partnership produced the eukaryotic cell. An essential element of their thesis is that mitochondria came early and were a prerequisite to the evolution of eukaryotic organization. I must reserve judgment about parts of this proposal: I have always doubted that merger of prokaryotic cells, and would note that an early acquisition of mitochondria is entirely consistent with Carl Woese's original view that the progenitor of the eukaryotes represents one of the primary lines of cellular descent. The primordial community of prokaryotes would surely have had a niche for some sort of primitive scavenger or predator (presumably related to the Archaea), which would make the perfect host for endosymbionts (10); we proud eukaryotes must acknowledge humble ancestry! Be that as it may (and we will probably never know for sure), the thesis that mitochondria came early and were required for the evolution of the full eukaryotic order has, to my ears, the ring of an important truth.

Let me not leave those evolvable eukaryotes without noting that the role of energetics in their emergence was permissive, not prescriptive: a generous supply of energy made great things possible, but mandated none of them. We are still obliged to specify the features of eukaryotic organization that natural selection favored, and the molecular inventions that marked the way. That task has not yet been fully accomplished, but the argument that predation based on phagocytosis is key to the rise of the eukaryotes (6,7) seems to me sound. Prokaryotes, by contrast, are constrained by limitations on the energy available to the genome, which bent their evolution into quite different channels. In their case, natural selection favored small and spare cells with streamlined genomes, rapid reproduction, little superfluous DNA and tightly disciplined regulation of gene expression. Marching under the banner Small is Beautiful they flourished, multiplied, and inherited the earth. From any point of view except that of a eukaryotic chauvinist, it's still a prokaryotic world.

References

1. **Kirschner M, Gerhart J.** 1998. Evolvability. *Proc Natl Acad Sci USA* **95:**8420–8427.
2. **Pigliucci M.** 2008. Is evolvability evolvable? *Nat Rev Genet* **9:**75–82.
3. **Lonhienne TG, Sagulenko E, Webb RI, Lee KC, Franke J, Devos DP, Nouwens A, Carroll BJ, Fuerst JA.** 2010. Endocytosis-like protein uptake in the bacterium *Gemmata obscuriglobus*. *Proc Natl Acad Sci USA* **107:**12883–12888.
4. **Lane N, Martin W.** 2010. The energetics of genome complexity. *Nature* **467:**929–934.
5. **Allen JF.** 2003. The function of genomes in bioenergetic organelles. *Philos Trans R Soc Lond B Biol Sci* **358:**19–37, discussion 37–38.
6. **Cavalier-Smith T.** 2002. Internat. *J. Sys. Evol. Microbiol.* **52**: 297–354.
7. **de Duve C.** 2007. The origin of eukaryotes: a reappraisal. *Nat Rev Genet* **8:**395–403.
8. **Embley TM, Martin W.** 2006. Eukaryotic evolution, changes and challenges. *Nature* **440:**623–630.
9. **Cox CJ, Foster PG, Hirt RP, Harris SR, Embley TM.** 2008. The archaebacterial origin of eukaryotes. *Proc Natl Acad Sci USA* **105:**20356–20361.
10. **Kurland CG, Collins LJ, Penny D.** 2006. Genomics and the irreducible nature of eukaryote cells. *Science* **312:**1011–1014.

January 16, 2011

bit.ly/1QNSuB9

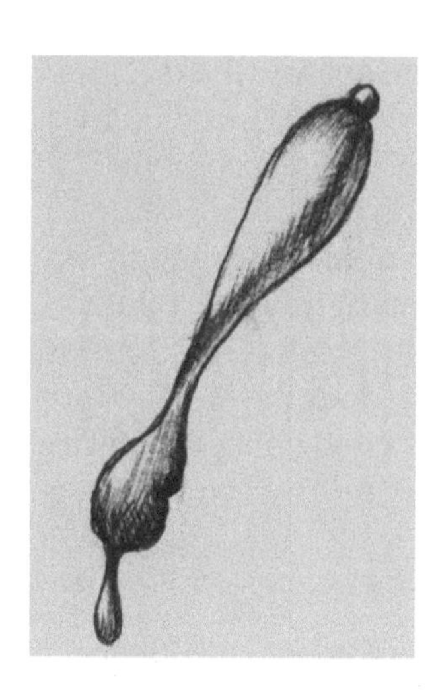

59

The Road to Microbial Endocrinology

by Mark Lyte

The emergence of the field of microbial endocrinology did not follow a straight course, nor was the destination intuitively obvious at the start. Like many research paths before it, the theoretical underpinnings of the field only emerged when it became obvious that the only explanation for the facts was a new awareness, in this case the realization that microorganisms are more interactive with the host than previously recognized.

The development of microbial endocrinology, the intersection of microbiology and neurophysiology, has its roots in the field of stress research. In the 1980s, there was growing recognition that host neurophysiology, specifically the elaboration and production of a panoply of various neuroendocrine hormones/neurotransmitters, modulated certain immunological phenomena. For example, the stress-related family of neurohormones, the catecholamines, could influence the production of antibodies against specific antigens. The direct interaction of neuroendocrine hormones on immune cell function was being examined from a mechanistic standpoint (i.e., direct examination of hormones and immune cells in an *in vitro* system). But beyond that it became apparent that the nervous system as a whole regulates immunity. The involvement of the brain and associated systems in the direct control of immune responsiveness was led by Ader and Cohen, who established the field of psychoneuroimmunology (1). It was in this field that I found myself as a newly minted assistant professor with my first NIH grant.

The primary objective of my grant was to examine the ability of social conflict stress in mice to modulate immune responsiveness. Social conflict stress is an ethologically relevant means to induce stress in animals and has been extensively used to elucidate pathways underlying stress-induced anxiety and for the pharmacological design of "anxiolytic" (stress-reducing) drugs. Although I had extensive experience in microbiology as a medical board certified clinical laboratory scientist, at that time I considered myself primarily an immunologist. As such, my thinking was largely shaped by the need to examine the mechanisms by which stress could influence immunity. My initial results were promising, starting with publication of a series of papers in the late 1980s and early 1990s showing that social conflict stress could indeed modulate immunity (2). Our results suggested that acute stress should be beneficial for the host by dramatically elevating the phagocytic capacity of macrophages.

What to do next? That stress elevated first-line immune defense suggested that it protected against infection. The obvious experiment—not envisioned in my original NIH grant—was to challenge the animal with an infectious microorganism and then subject it to the stress of social conflict. The results of these experiments presented the paradox that led to microbial endocrinology. What I observed in 1991 was that animals that had been given an oral infectious challenge had *decreased* rates of survival as compared to similarly challenged animals which were not stressed. On the face of it, this didn't make sense. How could animals with increased capacity to phagocytize bacteria be dying at a faster rate than the controls?

In order to resolve this seeming paradox, I took a fresh look at the underlying premise of my otherwise straight-arrow approach, i.e., that stress affects immunity. It had been known from decades of research that host neuroendocrine hormones/neurotransmitters play a large role in shaping the immune system's response to infectious challenge. But, could elements of the stress response *directly* influence the infectious microorganism as well? Such a consideration in no way obviated the role of immunity in the response to an infectious microorganism. On the contrary, if the microbes could respond to the neuroendocrine outflow resulting from the host's stress response, this meant that the microbes were *interactive* players in determining the outcome. Thus, the environment of the stressed host was much more complex, and most importantly, more interactive, than previously thought. Put concisely, I went from the more simple scheme:

STRESS	IMMUNITY	SYMPTOMS/DISEASE

to the more complex and interactive one:

	IMMUNITY	
STRESS		SYMPTOMS/DISEASE
	MICROBE	

A number of questions arose as I began to consider the microorganism as also capable of responding to the host's stress response. First, why should stress be suppressive? It didn't make any sense that when an animal is presented with an infectious challenge, such as occurs all the time in nature (think about being bit by someone in a fight), it should suppress its immune system. Well, my macrophage results showed that indeed stress was not suppressive; it was actually immune enhancing. However, I was still left with the initial problem of why stressed infected animals succumb faster to an infectious challenge. The answer appeared when I considered that the microorganism were just as capable of responding to host stress as the immune system.

In other words, what about the poor microorganism that is living happily in a culture dish with all the nutrients it needs and now finds itself stuffed into a mouse and having to contend with all the ensuing perils from both host and resident microbes? Isn't the infecting bug also being stressed? (Or course, I acknowledge the extensive literature about how alterations in environmental conditions affect microbes.) And concomitant with that was the realization that disease is often the result of exposure to a very low infectious dose, sometimes on the order of 1–10 CFU, such as in the case of *E. coli* O157:H7. At such a low dose in the gut, how would the immune system detect a few bugs? Having an enhanced immune system due to stress may be fine, and makes evolutionary sense, but it is the microbe that must first respond to the stress within the host. If so, what type of host environmental signals would it recognize so as to institute the physiological responses needed for survival and proliferation?

Stress has been described as the most ambiguous term in biology. Nonetheless, the neurophysiological outputs of stress, namely the neurochemical mediators, are well understood. For social conflict stressed mice, there would be a commensurate large release of the fight-or-flight neurohormones from the catecholamine family, namely norepinephrine and epinephrine. I performed the simplest of experiments. I simply mixed the infectious bacterium I had used for the challenge studies with norepinephrine and epinephrine in separate Petri dishes and set them in the incubator. Following overnight incubation, the bacteria in the norepinephrine-containing dish had grown to fill the entire dish while the epinephrine- and control-containing dishes evidenced hardly any growth. I had my answer—*micromolar amounts of a neurochemical could enhance the growth of a bacterium*–and the start of a new field which I termed microbial endocrinology (3, 4). Microorganisms were not simply "dumb bugs" but they actively recognize host neuroendocrine hormones.

The inception of microbial endocrinology was guided by the design of the media for *in vitro* testing of microbial-neuroendocrine interactions. Since environmental factors (i.e. neuroendocrine hormones) could influence microbial survival, why shouldn't the test media also be considered an environmental factor? In other words, the context in which the microbes were to be evaluated could prove to be as important as the factor to be studied. For these reasons, I chose to use iron-restricted serum- (or plasma-) based media. Also, they are difficult systems for bacteria to grow in and they more closely mimic the *in vivo* milieu. This approach has over the years led to some interesting exchanges, with one particularly standing out in my mind where a session chair asked in incredulity why I would choose such a difficult medium to evaluate neuroendocrine hormones when such good rich media such as LB, BHI, etc., exist. My answer was simple: because we don't have LB and BHI floating through our veins. If one seeks to evaluate the effects of a purported host factor on microbial physiology, one needs to do it in an environment close to the one found in the host.

And finally, I always caution that if you think that you are the first to discover something, you may not have looked hard enough at the literature, and specifically that which existed before the advent of scientific databases such as PubMed. A couple of years after my initial publication, I found that work done in France in 1930 had essentially observed the same phenomena as I reported in 1991 (for review of past studies see [5]). From there I followed a trail through the 1950s where numerous reports documented that neuroendocrine hormones influence microbial growth. When it came to an explanation, none proposed that bacteria could actively recognize host hormones; bugs were simply too "dumb" and the answer had to be exclusively on the host side.

I have not touched upon the other direction: microbes produce some of the same neuroendocrine hormones we possess in our brains, which means that not only can we influence microbes through production of our hormones, but also that the microbes can potentially influence us (6). I have been heartened by the growing number of reports of microbial-neuroendocrine hormone interactions, an indication that the theory of microbial endocrinology will continue to be examined and may contribute to a better understanding of the ways in which microbes interface not only with vertebrates, but also with any organism (such as plants) that produces neuroendocrine hormones.

Mark Lyte is Professor at the Center for Immunotherapeutic Research, School of Pharmacy, Texas Tech University Health Sciences Center.

References

1. **Ader R, Cohen N, Felten D.** 1995. Psychoneuroimmunology: interactions between the nervous system and the immune system. *Lancet*. **345**(8942):99-103.
2. **Lyte M, Nelson SG, Thompson ML.** 1990. Innate and adaptive immune responses in a social conflict paradigm. *Clin Immunol Immunopathol*, **57**(1):137-147.
3. **Lyte M.** 1993. The role of microbial endocrinology in infectious disease. *J Endocrinol*, **137**(3):343-345.

4. **Lyte M, Ernst S.** 1992. Catecholamine induced growth of gram negative bacteria. *Life Sci*, **50**(3):203-212.
5. **Lyte M.** 2004. Microbial endocrinology and infectious disease in the 21st century. *Trends Microbiol*, **12**(1):14-20.
6. **Lyte M.** 2011: Probiotics function mechanistically as delivery vehicles for neuroactive compounds: Microbial endocrinology in the design and use of probiotics. *BioEssays: news and reviews in molecular, cellular and developmental biology*, **33**(8):574-581.

August 27, 2012

bit.ly/1LHEjNP

by Elio

What's your guess as to the number of different kinds of viruses that can infect a single species on average?

September 15, 2011

bit.ly/1W1NUk7

60

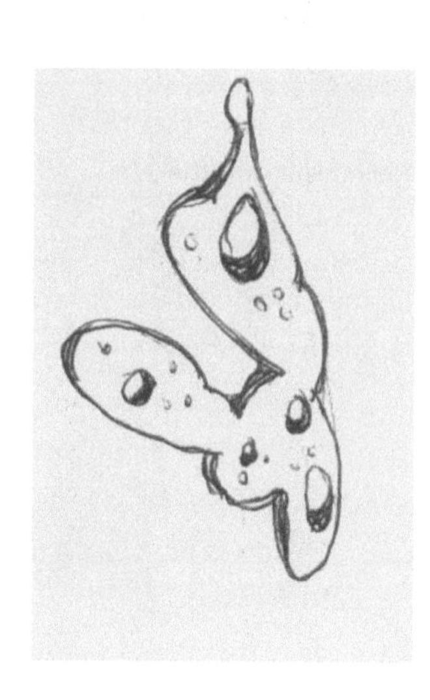

Begetting the Eukarya: An Unexpected Light

by Franklin M. Harold

Concerning the origin of eukaryotic cells, much has been written but almost everything remains to be settled. No one disputes that mitochondria derive from free-living bacteria that established an intimate symbiotic relationship with a host of some kind and progressively turned into organelles, workhorses of metabolism, and a hallmark of eukaryotic organization. But consensus ends here; the nature of that host, the timing and circumstances of the partnership, and its role in generating the conspicuous complexity of eukaryotic cells all remain elusive and entangled in controversy. In a comprehensive review of the subject published in 2006, Martin Embley and William Martin (6) concluded bleakly that: "*the evolutionary gap between prokaryotes and eukaryotes is now deeper, and the nature of the host that acquired the mitochondrion more obscure, than ever before*." No one will claim that the darkness has been lifted, but just in the past few years this intractable subject has begun to appear in a fresh light.

A good place to start is an article by Miklòs Müller and nine coauthors in the June issue of MMBR (15). Unless you are in the business, a meaty review of energy metabolism in anaerobic eukaryotes may not set your pulse racing; but if you are curious about the origin of mitochondria and their place in the history of the eukaryotic cell, you will find here an evidence-based summary of where that inquiry stands. For present purposes, let me highlight just three points from that article.

1. The diversity of eukaryotic energy metabolism is extremely

limited, much narrower than that of prokaryotes. Most eukaryotes live by mitochondrial respiration, with oxygen as the ultimate electron acceptor, and generate the bulk of their ATP by a chemiosmotic proton circulation coupled to an ATP synthase. This pathway is supplemented by fermentative reactions in the cytoplasm linked to the classical Embden-Meyerhof scheme of glycolysis. Anaerobic eukaryotes lacking conventional respiration are widespread, represented by over a thousand species. In these, ATP is produced by substrate-level phosphorylation, chiefly by the decarboxylation of pyruvate, which is commonly associated with mitochondria or their evolutionary relatives. The pathways of eukaryotic energy metabolism, including the anaerobes, are all very much the same; the participating enzymes represent a sub-set of a eukaryotic complement that was probably already present in LECA, the last eukaryotic common ancestor.

2. All contemporary mitochondria appear to share a common ancestry from among the alpha-proteobacteria. The closest living relatives of that puissant symbiont are thought to be facultative anaerobes such as *Rhodospirillum rubrum* and *Rhodobacter capsulatum*, capable of growing by respiration, fermentation, or even photosynthesis, as circumstances require. The genomic evidence suggests that all extant mitochondria descended from a single episode of symbiosis, a unique event in the history of life.

3. All eukaryotes that have been examined in sufficient detail, a sample that now includes representatives of all six supergroups, contain either standard mitochondria or organelles related to mitochondria by descent. That tribe now includes anaerobic mitochondria (whose respiratory chain terminates in an endogenously produced electron acceptor such as fumarate), a variety of hydrogenosomes (energy-generating organelles that lack respiratory chains, produce ATP by substrate-level phosphorylation, and typically generate hydrogen gas as a by-product), and mitosomes

(highly reduced, membrane-bound compartments that have no role in energy transduction). The strong implication is that mitochondria are not an optional accessory to eukaryotic cells but part of their basic fabric.

Taken together, these findings (and others omitted from this condensed summary) support a coherent thesis, one that departs quite radically from the "standard model" (5). Until about a decade ago, the general presumption was that the essential features of eukaryotic organization evolved autogenously in a proto-eukaryotic lineage, without contributions from other organisms. The symbionts that eventually became mitochondria would have been acquired, probably by phagocytosis, by a host that already possessed the essential characteristics of eukaryotic cells, including a nucleus, endomembranes, and a cytoskeleton (1, 4, 11). This entity has been dubbed the "primitive phagocyte." Müller et al., and many others (see below), advocate a different take on the genesis of the eukaryotic order. At an early stage in the evolution of what would become the eukaryotic cell, a facultative bacterium akin

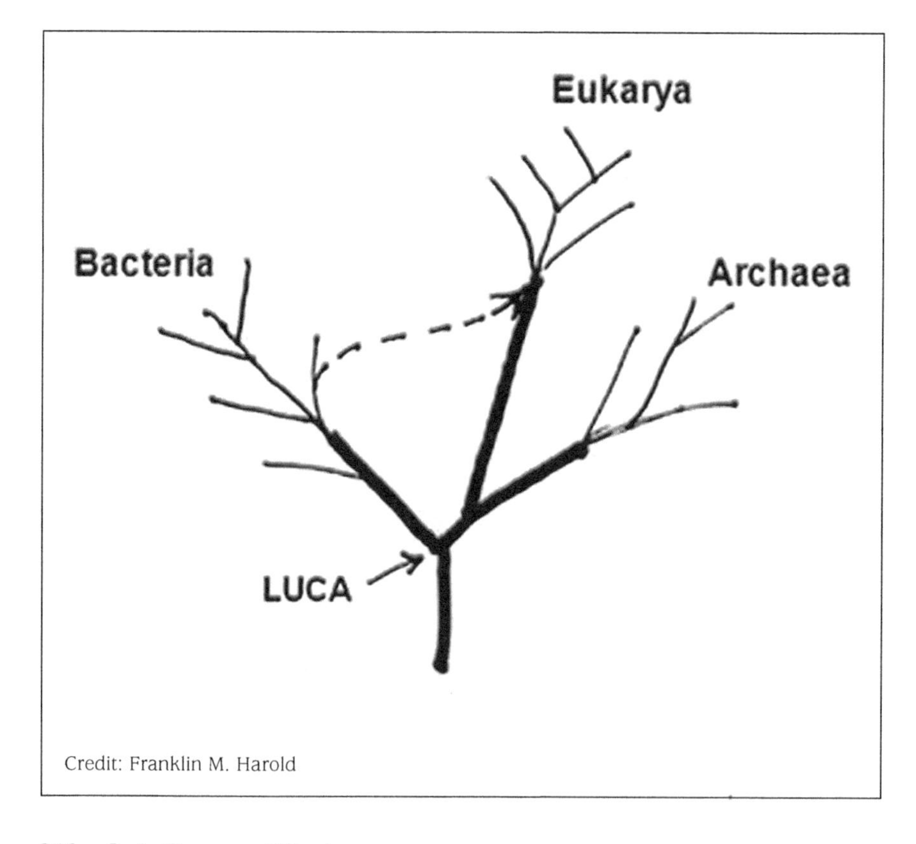

Credit: Franklin M. Harold

to today's alpha-proteobacteria forged an intimate association with a host whose nature is left open (but see below). By then, the diversification of the bacteria must have been well underway. Oxygen had begun to accumulate, at least locally, but the atmosphere and the oceans were still largely anaerobic (and would long remain so). A date around two billion years ago seems a fair guess. That was a fateful encounter, quite possibly a unique event. The symbiont contributed many genes to the consortium, and set it on track to evolve greater complexity, autonomy, and sophistication than anything known in the prokaryotic universe.

What was it about symbiosis that made so huge a difference? As outlined in an earlier contribution to this blog (21), I have been impressed by the thesis of Lane and Martin (12) that it's basically about energetics: thanks to its many compartmentalized energy packs, derived from the symbiont, the consortium could mobilize far more energy per gene than could a prokaryotic cell. Energy galore supported expansion of the genome, which in turn underpinned rising size and complexity. So we have two very

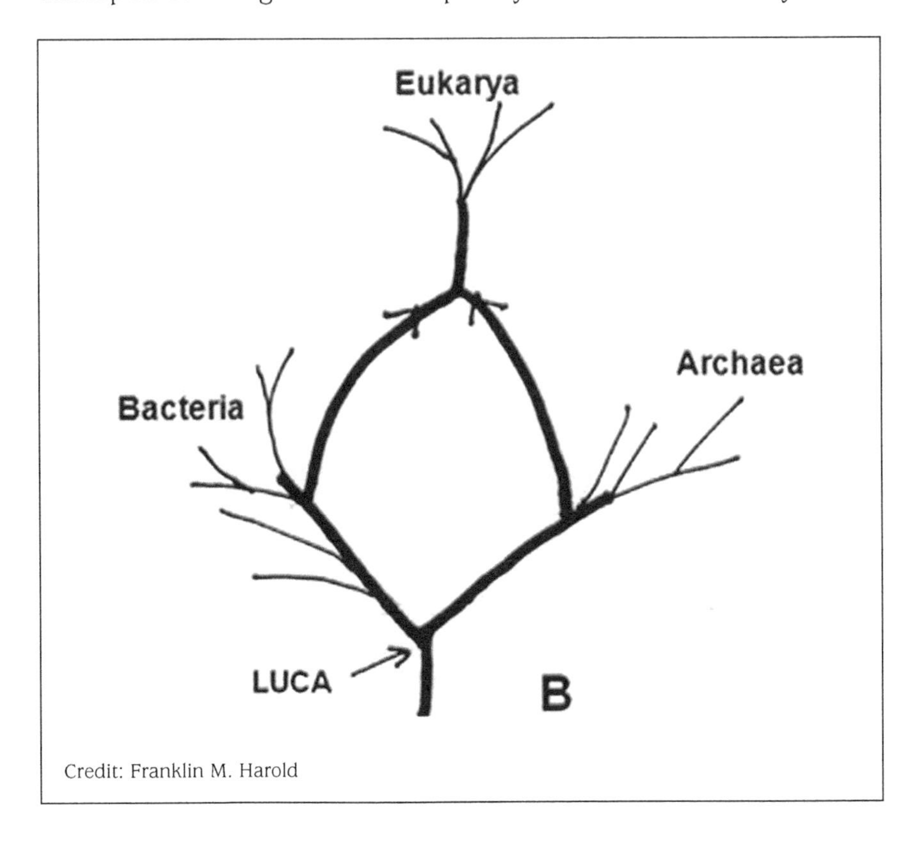

Credit: Franklin M. Harold

different narratives of the origin of mitochondria, and of the cells that harbor them. In one, symbionts come late into advanced proto-eukaryotes; in the other, they come early and help shape the eukaryotic order.

It takes at least two to make a relationship. Who, then, was the partner that hosted the precursor of mitochondria? That issue continues to bedevil the entire field of cell evolution, with advocates and evidence to support almost every conceivable answer.

Carl Woese and colleagues (16,18) taught us all that the nuclear lineage of eukaryotes is one of three primary lines of descent, a sister-clade to the Archaea, whose roots reach back almost as far as the last common ancestor of all cellular life. Thomas Cavalier-Smith has articulated and tenaciously defended the thesis that the elusive host originated from the Eubacteria, undergoing a series of transfigurations that eventually gave rise to both Archaebacteria and to the proto-eukaryotic phagocyte. And molecular biologists increasingly favor the proposition that the eukaryotic line is fundamentally chimeric, originating in a symbiotic association between two prokaryotic cells, one bacterial and the other archaeal. Depending on whom you read, that may have happened once or on several occasions, with the archaeon variously the donor of the consortium's genetic core, its nucleus, much of the cytoplasm or all of the above. To cite but one much-discussed example, William Martin and Miklòs Müller (14) proposed a syntrophic association in which a facultative anaerobic eubacterium, producing hydrogen gas, sustained a methanogenic archaebacterium; they also spelled out a sequence of transitions that would have turned that partnership into a eukaryotic cell. A decade ago, one expected the rising torrent of genetic and phylogenomic evidence to render a clean verdict. That did not happen, leaving a palpable sense of frustration (8).

Controversy is the lifeblood of science, and sometimes its bane too. But scientific disputations are seldom battles to the death. More often than not, they find resolution in a synthesis that draws on the insights of several protagonists but transcends them all, and that may be happening now with regard to the elusive eukaryotic host. Movement begins with the provocative finding from Martin Embley's laboratory (3) that the phylogenomic relationship between eukaryotes and prokaryotes depends on the particular statistical model employed to analyze it. Ribosomal RNA data have long been the mainstay of the conventional triple-stemmed tree of life, with the Eukarya as one of the primary lineages (Fig. A). But when the sequences are interrogated using the latest and most sophisticated procedures, they show something quite different: data from both rRNA and highly conserved proteins indicate that eukaryotes branch from within the Archaea, and are specifically related to the phylum Crenarchaeota. A similar conclusion was reached by Eugene Koonin's team (20): they have eukaryotes arising as an early branch of the Archaea, an unknown and possibly extinct lineage that diverged prior to the radiation of Archaea into their contemporary forms.

Such claims will surprise many but not James Lake, who first inferred a specific relationship between eukaryotes and "Eocytes" (= Crenarchaeota) a quarter of a century ago and has maintained that position ever since (17). His "ring of life" (Fig. B) has long been overshadowed by the three-domain tree of life, and if it rested solely on phylogenomic analysis, it would still be viewed with some suspicion. (My generation, at least, understands that statistics are like a bikini: what they reveal is interesting, what they conceal is vital.) But all of a sudden, confirmation has begun to pop up from other quarters, too. Most notable is the discovery of true orthologs of actin, apparently involved in cell morphogenesis, in both *Thermoproteales* (an order of Crenarchaeota) and in *Korarchaeota* (7,9), both deep branches of the archaeal bush. Orthologs of tubulin have turned up in *Nitrosoarchaeum*, a member of the novel phylum Thaumarchaeota, which is another ancient branch (19). Just to confuse matters, tubulins also occur in the bacterial phylum *Prosthetobacter*. And Crenarchaeota possess a membrane-remodeling system related to the ESCRT-III system of eukaryotes (13). Taken together, the data suggest that the eukaryotic lineage branched off early in the evolution of the Archaea, prior to the appearance of the current lines, and that they retain many traces of that common ancestry.

Don't look now, but just possibly a glimpse of the elusive host is taking shape: neither a classical primitive phagocyte nor a standard prokaryote, but something with features of both. A complex cytoskeleton, with the capacity to form membrane protrusions, hints at a novel way to make a living. Like Gaspar Jèkely (10), I like to imagine that the primordial microbial mats and biofilms offered a niche to primitive scavengers and predators that felt no need to compete with the prokaryotic primary producers; instead, they ate them. Sometimes—conceivably, just once—the prey evaded digestion and established residence. The host contributed the genetic core, the cytoskeleton, ATPases, and probably much else; the symbiont offered genes and a glorious future untrammeled by energy shortages. Many loose ends remain to be tucked in. Whence came signature genes found only in eukaryotes? How does the metabolism of eukaryotic cells relate to that of the hypothetical scavenger? Why is the lipid composition of eukaryotic plasma membranes akin to that of Bacteria rather than Archaea? Even so, it may be the case that we are watching the genesis of the Eukarya turn from a mystery into a puzzle.

For microbiologists preoccupied with practical matters, the begetting of the Eukarya will seem a nebulous subject best contemplated over a beer. But this arcane issue impinges on one of the conceptual foundations of microbiology, namely the tree of life with its three ancestral domains (Fig A). The narrative sketched above challenges that iconic image in two ways. First, if the lineage of the eukaryotic host branched off within the Archaea, is it still appropriate to consider it a separate, independent line? Well, that will depend on the precise relationship of the eukaryotic ancestor to the last common ancestor of Archaea, a matter unlikely ever to be unambiguously resolved. To me, the key point is that, from its inception, the eukaryotic branch seems to have operated on a distinctive pattern of organization; if any grouping rates the designation of domain, it's the Eukarya. The second, and more serious concern, turns on the role of the symbiont. If symbionts came late into a pre-formed proto-eukaryotic host, to which they made only a minor contribution, the image of the tree remains apt. But if, as now appears, they came early and helped shape the very fabric of the eukaryotic cell, then the Eukarya are not a primary domain but a secondary, or derived one (Fig. B). Thus, the most fitting image to hold in mind is not a tree, but a pointed Gothic arch thrusting out of the prokaryotic underbrush towards a higher plane of life. *Viva Eukarya!*

Frank Harold is an affiliate professor in the Department of Microbiology, University of Washington Health Sciences Center. Now retired, he remains engaged with science as a writer and unlicensed philosopher.

References

1. **Cavalier-Smith T.** 2002. The neomuran origin of archaebacteria, the negibacterial root of the universal tree and bacterial megaclassification. *Int J Syst Evol Microbiol* **52:**7–76.
2. **Cavalier-Smith T.** 2010. Deep phylogeny, ancestral groups and the four ages of life. *Philos Trans R Soc Lond B Biol Sci* **365:**111–132.
3. **Cox CJ, Foster PG, Hirt RP, Harris SR, Embley TM.** 2008. The archaebacterial origin of eukaryotes. *Proc Natl Acad Sci USA* **105:**20356–20361.
4. **de Duve C.** 2007. The origin of eukaryotes: a reappraisal. *Nat Rev Genet* **8:**395–403.
5. **Doolittle WF.** 1999. Phylogenetic classification and the universal tree. *Science* **284:**2124–2129.
6. **Embley TM, Martin W.** 2006. Eukaryotic evolution, changes and challenges. *Nature* **440:**623–630.
7. **Ettema TJG, Lindås A-C, Bernander R.** 2011. An actin-based cytoskeleton in archaea. *Mol Microbiol* **80:**1052–1061.
8. **Gribaldo S, Poole AM, Daubin V, Forterre P, Brochier-Armanet C.** 2010. The origin of eukaryotes and their relationship with the Archaea: are we at a phylogenomic impasse? *Nat Rev Microbiol* **8:**743–752.
9. **Guy L, Ettema TJG.** 2011. The archaeal 'TACK' superphylum and the origin of eukaryotes. *Trends Microbiol* **19:**580–587.
10. **Harold FM.** Energetics of the Eukaryotic Edge. *Small Things Considered.* N.p., 16 Jan. 2011. Web. September 10, 2012
11. **Jékely G.** 2007. Origin of phagotrophic eukaryotes as social cheaters in microbial biofilms. *Biol Direct* **2:**3.
12. **Kurland CG, Collins LJ, Penny D.** 2006. Genomics and the irreducible nature of eukaryote cells. *Science* **312:**1011–1014.
13. **Lane N, Martin W.** 2010. The energetics of genome complexity. *Nature* **467:**929–934.*The energetics of genome complexiy.*
14. **Makarova KS, Yutin N, Bell SD, Koonin EV.** 2010. Evolution of diverse cell division and vesicle formation systems in Archaea. *Nat Rev Microbiol* **8:**731–741.*Evolution of diverse cell division and vesicle formation systems in Archaea.*
15. **Martin W, Müller M.** 1998. The hydrogen hypothesis for the first eukaryote. *Nature* **392:**37–41.*The hydrogen hypothesis for the first eukaryotes.*
16. **Müller M, Mentel M, van Hellemond JJ, Henze K, Woehle C, Gould SB, Yu RY, van der Giezen M, Tielens AG, Martin WF.** 2012. Biochemistry and evolution of anaerobic energy metabolism in eukaryotes. *Microbiol Mol Biol Rev* **76:**444–495.
17. **Pace NR.** 2009. Mapping the tree of life: progress and prospects. *Microbiol Mol Biol Rev* **73:**565–576.
18. **Simonson AB, Servin JA, Skophammer RG, Herbold CW, Rivera MC, Lake JA.** 2005. Decoding the genomic tree of life. *Proc Natl Acad Sci USA* **102**(Suppl 1)**:**6608–6613.

19. **Woese CR, Kandler O, Wheelis ML.** 1990. Towards a natural system of organisms: proposal for the domains Archaea, Bacteria, and Eucarya. *Proc Natl Acad Sci USA* **87:**4576–4579.

20. **Yutin N, Koonin EV.** 2012. Archaeal origin of tubulin. *Biol Direct* **7:**10.

21. **Yutin N, Makarova KS, Mekhedov SL, Wolf YI, Koonin EV.** 2008. The deep archaeal roots of eukaryotes. *Mol Biol Evol* **25:**1619–1630.

September 10, 2012

bit.ly/1M3Tmn5

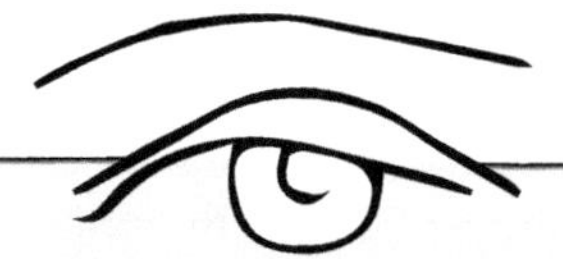

#24

by Elio

Why didn't mitochondria and chloroplasts cede all their DNA to the nucleus, as did most hydrogenosomes and certain other organelles?

November 19, 2007

bit.ly/1OPjd14

61

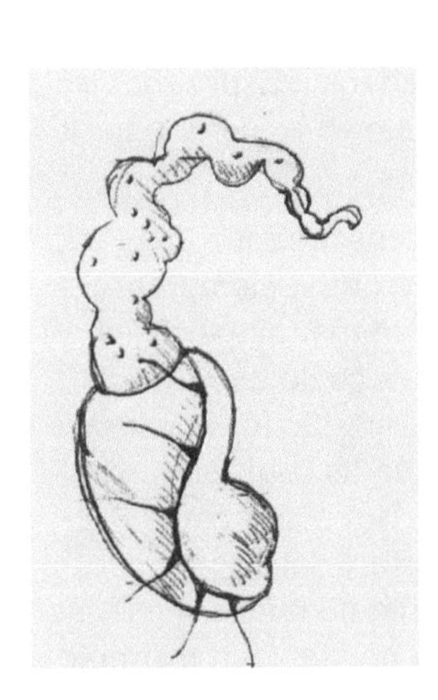

The Higgs Boson and Biology

by Elio

I approached several physicists, some turned biologists, to ask them for a brief comment on the topic: "*In the long run, what will the discovery of the Higgs boson do for biology?*" Their answers span the extremes from "nothing" to "everything."

Joe Incandela,
Dept. of Physics, University of California at Santa Barbara
(Joe was the head of the CMS team, one of the two groups that carried out the experiments at the CERN Large Hadron Collider that observed the Higgs boson)

The Higgs field is related to the state of the universe. Our universe is surprising in many respects because it appears to have a remarkable near-perfect balance of underlying characteristics that allows for everything we see to exist. We cannot yet explain this balance and it could be just random chance. With string theory it is hoped that we can explain a lot about why the universe has the state that it has but it could turn out that there are in fact a huge number of possible universes, maybe all of them exist, and most of them by far do not come anywhere near to having the characteristics of our universe. So our universe could be the extremely rare exception that is balanced in this way and it is only in this balance that there can be life. The Higgs, which is a part of this picture, is then simply a byproduct of the fact that we are in the rare case of a universe where we can exist and ponder these questions. I think it is safe to say that we hope that we can eventually explain why the universe has the characteristics that it has and that it is not just random chance. Meanwhile, getting back to the Higgs...

Without a Higgs field (of which the Higgs bosons are the corresponding quanta) it is believed that none of the elementary particles would have mass. This does not mean that no particles would have mass because elementary particles could bind to one another and in so doing create a massive particle. So for instance, protons and neutrons which are made of 3 quarks could still exist but the electron would remain massless. The latter is of course of very great consequence because it means there would no longer be atoms and this of course means there would be no elements, no compounds, no structure of any kind, and so no life. So the simple relation of Higgs to biology is that without it there is no biology.

Suckjoon Jun,
Center for Theoretical Biological Physics,
University of California at San Diego

It has been many years since my student days when I naively thought that I would become a theoretical particle physicist. I reckon physicists' reaction to the news about the Higgs boson is like an excitement rather than a surprise. It's like filling one glaring hole in the periodic table with the element "X" that we know must exist—that is, if you believe in chemistry (or quantum mechanics).

I am not really sure how all this will do for biology. On the one hand, I am personally curious to know why biological systems are the way they are, and whether fundamental physical principles matter in my study. On the other hand, I hear this lingering voice of John Maynard-Smith in *The Major Transitions in Evolution*: "... *This book is about how and why this complexity has increased in the course of evolution. The increase has been neither universal nor inevitable....*"

Stefan Klumpp,
Max Planck Institute for Colloids and Interfaces, Potsdam

I have three answers to the question.

1. Nothing.

2. Maybe now all the smart people who have been searching

for the Higgs boson move on, turn to something else, and some of them end up making great contributions to biology.

3. On a more serious note, I imagine two things the search and discovery can contribute to other fields (such as biology). The first is an appreciation for theory. A theoretical prediction being confirmed by a decades-long experimental endeavor, just imagine that! The second contribution, more one of high-energy physics in general than of the search for the Higgs boson specifically, is the technology developed on the way (the use of synchrotron radiation from particle accelerators for structural biology is probably the most direct example).

David Nelson,
Dept. of Physics, Harvard

The Higgs boson is an impressive discovery in fundamental science, but I don't think it will affect day-to-day biology, chemistry or even physics very much.

Experimental particle physics has now burrowed down to length scales so incredibly small that its new scientific discoveries are largely insulated from what happens at length scales larger than an Angstrom, i.e., the size of an atom. Finding the Higgs boson is bit like completing the row of trans-uranium elements in the periodic table. Most of those exotic elements are quite unstable, and so is the Higgs boson. According to *The Economist* magazine, the last really "useful" particle discovered in high energy physics was the neutron in 1932!

Of course, judgments of this kind should be viewed with caution—perhaps the Higgs boson will lead to discoveries that shed light on profound cosmological questions such as the origin of dark matter or dark energy.

Rob Phillips,
Applied Physics, Biochemistry and Molecular Biophysics, Caltech

The original predictions about this Higgs boson were made in the 1960s, nearly 50 years ago. Such predictions are also a key part (to my understanding) of the so-called "standard model" of particle physics. There is a very deep lesson here. First, that the job of theorists is to make polarizing predictions about experiments that have not yet been done. I am strongly disturbed by the attitude of journals that insist on "data" to go along with theoretical predictions. In fact, the most exciting and interesting predictions are those for which there is no data. I just got a request to be a guest editor this morning and here is the last line: "*While experimental validation is not necessarily required for publication, we do, however, seek substantial evidence for conclusions.*"

The Higgs boson story shows why that is so wrong. It is beautiful that over such a long period of time, a community could come together in their conviction that one should make bold predictions and then test them. Indeed, it always makes me wonder why the word "speculative" is so pejorative and thrown around as though it is a damning indictment of someone's science. So, to sum up, my belief is that forward progress in science depends largely on the interplay between bold, polarizing predictions and careful measurements specifically targeted to test those predictions. In a better world where all scientists saw themselves as part of a common cause to better understand nature, biology would take away inspiration from the kind of rich interplay between theory and experiment that is evidenced by the Higgs story.

September 13, 2012

bit.ly/1LHEmt1

62

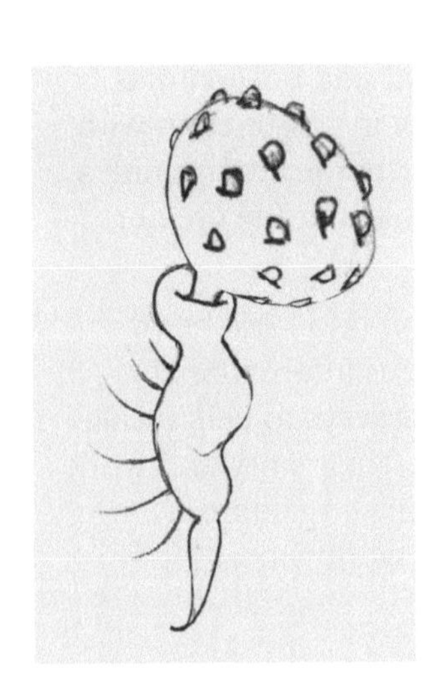

The Power of Fungal Genetics

Cassava for Food Security and Sustainability in Colombia

by Chris Condayan
Peter Geoghan

ASM's *Cultures* magazine traveled to Colombia for its recent food issue to capture its first digital multimedia story.

ASM traveled to Colombia to speak with and film the researchers behind an innovative biotechnology project that is producing exciting results. The international Swiss–Colombian collaborative research team from the University of Lausanne (Switzerland), the Universidad Nacional de Colombia, and the Universidad de la Salle (Utopia campus) has been working to create and test novel strains of arbuscular mycorrhizal fungi (AMF) to improve cassava production.

AMF forms symbiotic relationships with the majority of the world's plant species, including cassava and other major food security crops. By colonizing internal structures within the plant and extending its root system, AMF transports nutrients such as phosphate to the plants from inaccessible areas and sources in the soil. In exchange, the plant provides carbon to AMF species that have colonized the plant.

The research team's studies show that, with the inoculation of certain AMF strains, only half of the necessary phosphate amendments are needed in nutrient-poor tropical soil to produce an equal or greater amount of cassava yield. On a large scale, this technology could potentially provide a more sustainable approach to resource management, allow small shareholder farmers to reduce their input costs, and help create a food secure future for many. In fact, an early model for this success is already being realized by graduates of the

Utopia campus, all of whom come from conflict and post-conflict zones. By utilizing their education in agronomy in conjunction with this technology, they can begin rebuilding their home communities while ensuring a food secure future for Colombia and the greater global community.

We hope that you enjoy this *Cultures* production (available here [bit.ly/1NBd2w4]) and the incredible work of this international collaborative team of scientists who are finding ways to help tackle issues surrounding a major food security crop. Their story and work is multi-faceted, overcoming geographical distance to address an issue that affects people in over 100 countries worldwide.

About *Cultures*

Cultures magazine is a quarterly publication at the American Society for Microbiology that explores the nexus of science, policy, and global challenges we all share by bringing diverse voices to a common platform. Now in its second year, *Cultures* is being read by some 50,000 people in 160 countries, reaching young scientists in some of the most remote regions of the world. Past contributors to the magazine include President Jimmy Carter, Congresswoman Louise Slaughter, and CDC Director Tom Frieden. We are excited to announce that the first issue of this year includes the voices of Dr. Jane Goodall and other top scientists who provide guidance on creating food systems for the future. In addition, *Cultures* has produced its first ever video production, exploring food security through fungal genetics to improve cassava production in Colombia.

Chris Condayan is Video Producer for the American Society for Microbiology.

Credit: Charles Steck.

March 12, 2015

bit.ly/1NliItT

63

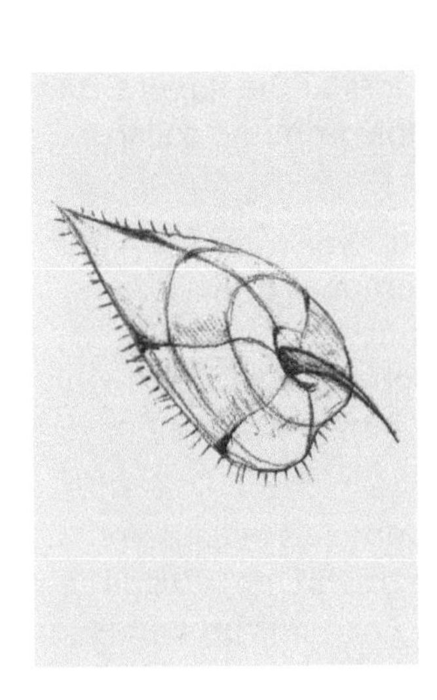

Why Is It So Hard to Make an HIV Vaccine?

by John Coffin

The answer lies in the unusual relationship of this particular virus with its host.

Searching for Achilles's Heel

By and large, viruses of humans and other animals interact with their hosts in one of two ways. After an initial period of rapid (often symptomatic) infection, some viruses (e.g., influenza) are cleared completely by the immune system. Others (e.g., many herpes viruses) establish a lifelong latent infection, only to reappear much later when conditions allow. In either case, the immune system effectively clears the virus from the host. Once established, the adaptive humoral response can prevent future infection with antigenically similar viruses. This is what makes vaccines work.

Relatively recently, we learned that some retroviruses, including HIV and its dozens of primate lentivirus relatives, have evolved to infect hosts other than their original ones. In the original hosts, the primary infection is usually fairly mild, but the virus is only partially cleared by the immune system. Later, virus replication and cell killing continue at the same pace for the life of the host, although at a lower level. Usually, this process has little or no effect on the host's lifespan, allowing the opportunity for transmission to naïve animals. When the virus jumps to a new host (such as the recent transfer of HIV-1 from chimpanzees to humans and of SIVmac from monkeys called sooty mangabeys to rhesus macaques), the process of infection is essentially the same, but eventually something goes awry. Infection leads to the slow, progressive loss of the target $CD4^+$ (helper) T cells, and thence the nearly inevitable immune collapse and death of the host.

Clearly, the benign relationship of primate retroviruses with their primary hosts is the product of a very long period of coevolution. In the process, the virus has evolved two features to avoid the immune response and therefore to make an effective vaccine highly elusive to produce. Both arms of the immune response, humoral (antibodies) and cellular are so affected.

The virus is almost completely invisible to antibodies because of several unusual features of its envelope glycoprotein (gp120/gp41)—the normal target for neutralizing antibodies. First, this protein is adorned with an extraordinary amount of carbohydrate moieties on its exposed portion. Since these are added by host enzymes, they are seen as "self" by the immune system. Second, these carbohydrates, being on the surface, hide the virus-specific portion,

My good friend and occasional collaborator, Jonathan Stoye, with a Tootsie Roll Pop. Jonathan likes chocolate, but being British, he does not realize that there is a piece of chocolate inside all that sugar. In this respect, the immune system can be thought of as "British", in that it does not react to the foreign protein sequences of HIV gp120 under the layer of carbohydrate.

just as a layer of sugar conceals the chocolate in a Tootsie Roll Pop (my favorite model). Third is the remarkable rate of variation of several exposed loops on gp120, which allows the virus to easily outrun the relatively feeble immune response that targets them. Finally, the binding site for the CD4 receptor, which must remain exposed and invariant, is hidden from antibodies by its location at the bottom of a pit. The virus has compensated for the relatively low binding affinity for the receptor by evolving a two-step entry mechanism, where receptor binding triggers rearrangement of the protein, exposing binding sites for the coreceptor (CCR5 or CXCR4), which can then interact with sufficient affinity to trigger the rearrangement necessary for membrane fusion. The region exposed by CD4 binding is capable of generating an antibody response, but its brief existence and location at the virion-cell interface prevent effective access by antibodies.

The cellular immune response (via cytotoxic T lymphocytes, or CTLs) plays a significant role in modulating the level of infected cells but is incapable of clearing them altogether for a couple of reasons. For one, as the number of infected cells declines, the number of CTLs also declines. For another, HIV replication targets the $CD4^+$ T helper cells—the very cells that are essential for the most effective CTL killing. This leaves only a relatively weak helper cell-independent response. The upshot is that even when antiviral therapy has reduced the number of productively infected cells by as much as 6 orders of magnitude, the immune system cannot clear the small residue of infected cells. If therapy is discontinued, even after many years of treatment, the infection always recovers to previous levels.

To make things worse, there is no evidence that natural infection with HIV can induce preventive immunity. Indeed, there are plenty of examples of successful superinfection of a previously infected individual, even by closely related strains. Nor is there any single proven case of complete clearance of virus from an HIV-infected individual, with or without treatment. Furthermore, to my knowledge, a successful vaccine that relies solely on a CTL response has never been developed for any virus. Viewed in this light, the failures in large-scale clinical trials of vaccine candidates designed to elicit either neutralizing antibodies or, most recently, CTLs, should hardly come as a surprise to anyone.

Given the obvious need for effective prevention to stem the AIDS pandemic, we must keep trying. Unfortunately, HIV has evolved into a niche whose very properties seem designed to thwart our attempts to turn the immune system against it. As an article of faith, we must believe that there is an Achilles' heel in the virus's sugary armor that we can exploit, but we haven't found it yet.

John Coffin is American Cancer Society Professor of Molecular Biology and Microbiology at Tufts University.

September 29, 2008

bit.ly/1Pw0wkD

#111

by Elio

Can you conceive a viable (microbial) cell that has no ribosomes?

July 31, 2014

bit.ly/1PvBzG4

64

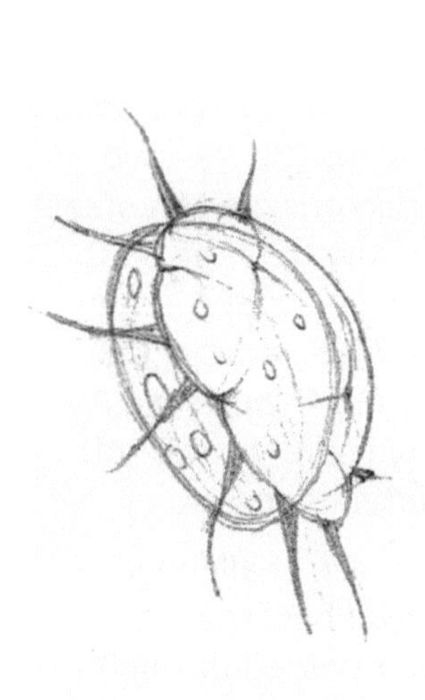

Measuring Up

by Jamie Henzy

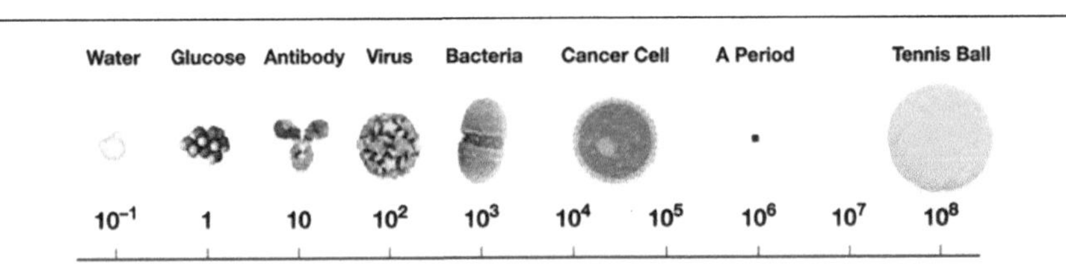

How do viruses and bacteria measure up? Adapted from National Cancer Institute Office of Cancer Genomics (OCG)

One could say that bacteria get quite a bit more attention than viruses here at STC. In fact, one could say that they hog the limelight, chewing the scenery and mugging for the camera while viruses, like stepchildren of a stage mom, hang backstage as understudies. So let's engage in an inclusive family activity, in which both bacteria and viruses can participate. How do they measure up to one another in size as they frolic together on the playground?

Bacteria generally range from 0.3 to 5 microns (one micrometer, or 10^{-6} meters), with *E. coli* a respectable 2 microns in length, and typically approaching 1 micron in width. Most eukaryotic cells would loom over a bacterium, being at least 10 to 20 times wider. But how about a virus stepsibling? Well, viruses of course are generally smaller. Take HIV, for example—a typical particle is about 0.1 microns in diameter, and somewhat spherical, so *E. coli* standing at its full height would be 20 times higher.

On the smaller end we find among bacteria the petite *Mycoplasma*, and among viruses the little runt, porcine circovirus (PCV). *Mycoplasma* measures 0.3 microns, and PCV, 0.02 microns. So Mycoplasma won't feel so small anymore when it towers over PCV at ~10 times its size. That is, until PCV's big sister Mimivirus steps in. One of the larger known viruses, she's a hefty (for viruses, that is) 0.5 microns, or about one and a half times the size of *Mycoplasma*. A virus larger than a bacterium! *Mycoplasma* would surely cower off with its flagellum between its legs—that is, if it had one.

Mycoplasma has recently received a boost to its size-challenged ego, however, with the discovery of the *ultra-small* bacteria. Researchers examined microbes from groundwater passed through a filter with a pore size of 200 nanometers and found bacteria with sizes as small as 0.22 by 0.18 microns—close to the assumed minimal sizes of live organisms worked out by a panel of the National Research Council in 1999.

Then there are the veritable monsters, such as the Gram-negative bacteria *Thiomargarita namibiensis*. These hulks are cocci that are typically 100 – 300 microns in diameter (that's 50 – 150 times larger than *E. coli*). Of course, the viruses have their own family freaks—the giant Pandoravirus and Pithiviruses, breaking the mold at 0.5 to 1.0 microns, and 1.5 microns, respectively. Okay, so not quite as large as *E. coli*, but more than three times the size of *Mycoplasma* ("little Mikey"?). Note, though, that *Thiomargarita* are Godzillas next to even these largest of viruses, 200 to 600 times larger.

Lastly, how do our subjects measure up to us? There are an estimated 10^{31} viruses on earth, and 10^{30} bacteria (viruses outnumber bacteria ten to one, ha!), and the total biomass of either group is far greater than that of all animals combined, prompting the question—who are the true stage hogs?

Jamie Henzy is a postdoctoral researcher in the lab of Welkin Johnson at Boston College, and an Associate Blogger for STC.

PART 7

Teaching Things

Some of us are irrepressible teachers, and we love to share tales from the front lines.

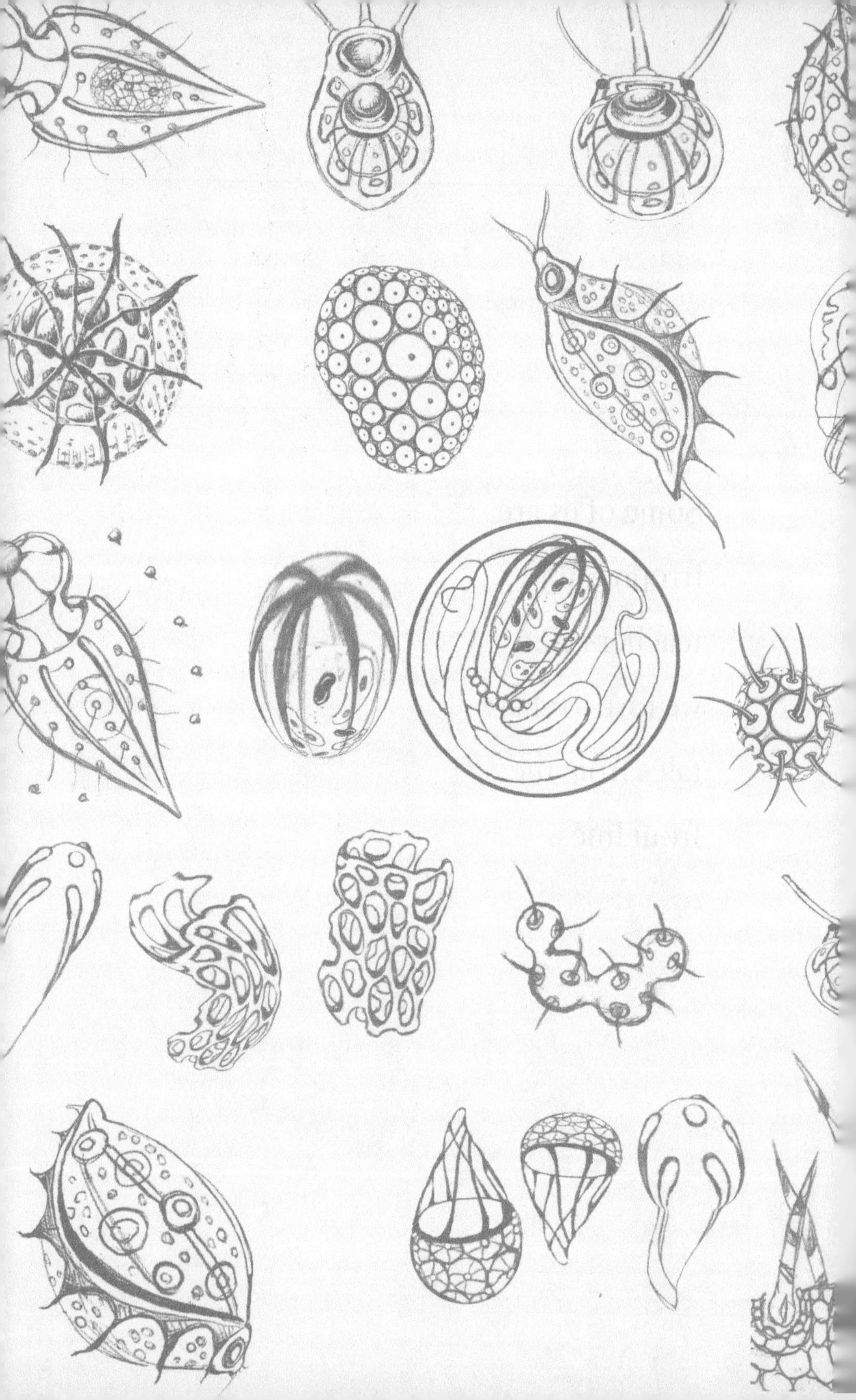

65 On Teaching

by Stanley Falkow

A graduate student came to my office recently to say that she was increasingly bothered by anxiety and the "terror" of having to speak at laboratory meetings. She had also learned a month ago that she was expected to lecture to a class organized by her mentor. The thought of having to lecture to 20 or so complete strangers had now led to sleepless nights, and she was physically ill as well as mentally distressed. She had been told that all of this could be "fixed" by a small dose of a beta-blocker or an anti-anxiety drug like Ativan. She had tried this but it did not help her enough, and she was now seeking my counsel because she had been told that I had written that I suffered from similar symptoms when I was a young scientist. Did I have a secret I could share? She confided in me that she was considering giving up her life long dream to become a "college professor." I gave her several pieces of advice and told her how I was first able to get some relief after my first attempt to lecture to medical students in 1966. It occurs to me that this student may not be alone and I have heard from a number of people that reading my "confession" that I had suffered from panic attacks early in my career had been helpful to them. So perhaps the readers of my favorite blog might find it helpful as well.

I had lived with anxiety and actual panic attacks, as well as the constant fear of having them, for over 10 years before I was asked to deliver a series of lectures across town from my job at Walter Reed to the sophomore medical class at Georgetown University. I had not yet established a truce with the demons living in my thoughts.

Yet, whenever I was faced with a decision to do something I feared, like accept an invitation to speak at a meeting or, as in this case, teach in front of an audience of 150 students, situations that were terrifying on the one hand but resonated with my ambition or scientific career, my first response was to say "*Yes, I'll do it!*" Indeed, I wanted to do it!! However, once alone with no defense against my self-generated fears, I would begin to plot how to avoid what I had agreed to do. This had become my MO for dealing with scientific meetings. I would accept and then the day before, I'd call and cancel because I was "sick" or gave some fabricated other excuse. But in this case, by accepting the invitation to teach at a local university, I understood that if I agreed, there would be no escape—nor did I want to escape. There was part of me that desperately wanted to "be on stage" and to take on the role of a teacher.

Those first lectures at Georgetown took place in the fall of 1966. I was absolutely terrified as I stood before the class. True, I had given a few research seminars before, but these were directed to small groups of people who by and large knew the basic subject matter. Not that I wasn't frightened when I gave research seminars. But, here in the classroom setting the terror I experienced was more global and more "dangerous" since I felt more exposed. As I began to lecture, I was sweating profusely, and my heart was pounding in my ears. I felt so light-headed that I thought I would, at any second, pass out. I was looking for something to hold on to in order to steady myself. There was a podium, but it was awkwardly placed to one side of the stage. So I stood on a stage in front of rising rows of seats occupied by a group of people not that much older than me who were by now looking quizzically at me and then one another. The faculty of the Department of Microbiology, led by Arthur Saz, were in the front row all waiting to see what I could or couldn't do. Medical students of that era were in class for hours at a time receiving detailed lectures illustrated by glass lantern slides one after another on subject matter that ranged from the dull to interesting to the exciting and even exultant. Which was I to be?

I began the lecture in a hesitating, nervous, much too low voice. "*Speak up,*" one of the students yelled. "*We can't hear you from up here.*"

I spoke louder but in a choking, tremulous voice. "*I want to discuss with you over the next several hours the biology, the epidemiology, and the nature of infection caused by bacteria that can infect the human intestinal tract.*"

I was certain that in any second I was going to faint from fright. I looked around for something to steady myself. There in the corner of the stage was an old wooden stool, the kind of furniture that one still sees in a bar or in school laboratory classrooms. I walked over to this wooden seat that was to be my haven of stability. I picked it up and carried it deliberately to the center of the stage and put it forcibly down. I then sat on it and faced the audience. I had not said a word since my faltering opening sentence. The audience was following my actions intently.

I continued, "*When dealing with potential gastrointestinal infection, one of the first things that a physician wants to know about is the patient's stool!*"

I swear to you that I had not intended the use of the word stool as a patient specimen to coincide with my successful search for something to sit upon that happened to be, well, a stool. The audience thought this was an intended play on words. They began to laugh. I looked at them. I was puzzled. I was even more horrified! What had I done?

"What's wrong?" I asked with a puzzled look on my face. They laughed louder at my feigned innocence. Faced with this laughter directed at me, I did what any man would do under the circumstances; I checked my fly to see if it was unzipped! Although I tried to do this discretely, all men will agree, it never is discrete enough. It did not escape the notice of the audience. Some people were now laughing really loudly. Fortunately, I suddenly realized what had happened and I smiled at them sheepishly with obvious relief. They laughed louder.

This unintended play on words that led to laughter relaxed the audience and me. I had never felt quite so at ease before in a lecture. So I began to deliberately look for humor to introduce into my lectures. When I told them about *Shigella* and dysentery I told them about my encounter with the monkey who had pelted me with feces. I exaggerated, I mimed. I played my own straight man and, relatively speaking, I relaxed. The lectures were a success. I had found by accident a means to relax while speaking in a group. I learned that the key to using humor in my lectures, at least for me, was to make myself the butt of the humor, never others. Since my subject matter was scatological, it was not difficult to get them to laugh. Shit is often funny.

I told the student who had sought my counsel that I did not suggest she become a stand-up comic. Rather, I suggested she try to find something in her lecture material that she felt most comfortable talking with others about and to make it her initial focus. "*Don't,*" I said, "*try to dazzle them with data and your in-depth knowledge of the minutia of the subject. Rather,*" I advised, "*talk to them as if you were explaining the material to your mother or, even better, your grandmother. It is what I try to do whether I teach, or give a scientific talk. I think it will relax you too.*"

Finally, I told her that I was not cured by this experience but it did provide me with a tool that got me through a tough time. I found, as well, that something that I had kept as a deep and dark secret about what I perceived was a fatal weakness in my character was not viewed that way by most people. I told her about the demons in my thoughts. Rather, they always would reassure me and try to help. Little by little I got better. And, besides, she had taken the first step by admitting to others what frightened her.

By the way, the demons in my mind still re-visit now and again, but I just tell them a bad joke and they give up and go away.

Stanley Falkow is Robert W. and Vivian K. Cahill Professor in Cancer Research, Emeritus, and Professor Emeritus of Microbiology and Immunology, Stanford University School of Medicine. He is also a past president of the ASM, where he has been a member for 61 years.

January 9, 2014

bit.ly/1RlmhkK

66

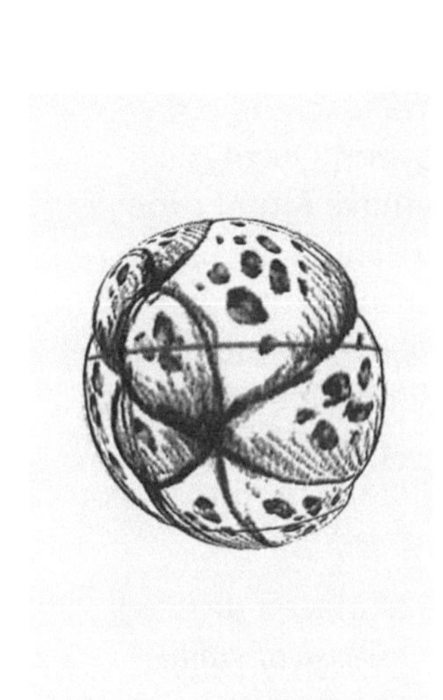

Hello Again, Metabolism!

by Amy Cheng Vollmer

Years ago, pathways of intermediary metabolism made up a significant portion of biochemistry and microbiology courses. Therein, students learned about interconversions and connections between pathways, and they could follow the carbons as they moved from acetate into the cholesterol molecule and many others. But the advent of exciting new methodologies—structural biology, recombinant DNA, molecular genetics, immunochemistry, probes, blots, microarrays, metagenomics, and more—crowded much of metabolism right off the syllabus. Given that teaching pathways could be dry and boring, faculty often elected to substitute more trendy and exciting topics instead. To be sure, they thought metabolism was important, but assumed "someone else must be teaching it."

The result is a generation of well-trained scientists who can clone or crystallize just about anything and can harvest bushels of data from vast microarrays. But once those gene names are converted into enzymes, they are not so adept at mapping their enzymatic steps into a coherent and integrated system of pathways. In fact, a survey I instigated at an IMAGE (Integrating Metabolism and Genomics) meeting in 2004 showed that the vast majority of individuals trained after the 1970s knew little about the pathways of photosynthesis or of amino acid, purine, and pyrimidine biosynthesis, nor did they think that needed to be taught at the undergraduate level.

So we now have faculty (in the assistant and associate ranks) who admit to me that they would have a tough time teaching pathways effectively because they don't know much beyond glycolysis and the Krebs cycle. Yet, in the past 5 years, I have found time and again that some of the most revealing presentations at the ASM meetings (and others) have shown that important signals in microbial processes are, in fact, small metabolites, and that the key enzymes are not specific to pathogenesis, development (e.g., sporulation), or differentiation (morphotypes), but rather they are the enzymes of the Krebs cycle or for key steps in nitrogen or phosphate metabolism. Imagine that!

In this era of metagenomics, metabolism has resurfaced dressed fashionably as metabolomics: *the study of the universe of small molecule metabolites that characterize a biological sample, usually one containing many different species,* e.g., the metabolome of the human gut or the cow's rumen. Somehow the profiles of small molecules reveal important aspects of the health of such ecosystems. So now we find faculty and students scrambling to teach and learn about primary and secondary metabolic pathways so that they can find their way through the vast databases that are being assembled from structures, pathways, regulatory networks, etc.

Metabolism is enjoying a renaissance in our curricula, and it is about time! There are vast secrets about biology to be revealed by the small molecules. Understanding how their levels rise and fall will take a careful study of the enzymatic pathways leading to and from them. So hello again, Metabolism! It's so nice to have you back where you belong.

Amy Cheng Vollmer is Professor and Department Chair of Microbiology, Swarthmore College, and President of the Waksman Foundation for Microbiology.

July 22, 2010

bit.ly/1W30WmN

67

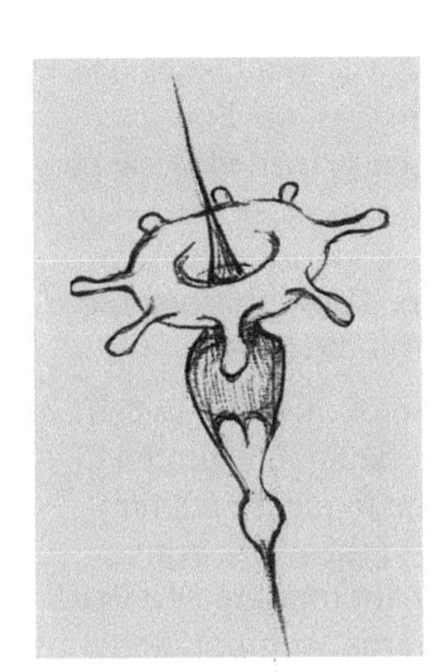

Why Medical Microbiology Is Not Like Stamp Collecting

by Elio

I started teaching microbiology to medical students in 1958, at a time when biomedical science was in its full ascendancy. Grant money was there practically for the asking, jobs were plentiful, universities and their medical schools were frenetically building up their science base. Not entirely surprising, the general feeling of young PhDs such as myself in basic sciences departments was that we were doing the medical students a favor by stooping to teach them. Maybe I exaggerate, but not by much. We had a long ways to go.

For decades, medical education had been driven by an influential study known as the 1910 Flexner Report. In general terms, the report declared that medical education must become seriously science-based in order for the practice of medicine to emerge from its dark ages. This was most laudable and enormously valuable, but in time the idea became subverted by the notion that the more science one could cram down the students' throats, the better. At times, this reached levels of utter mindlessness. A friend told me that in the early '50s, when he was a medical student at Harvard, a biochemistry professor wrote the amino acid sequence of insulin on the board, it having been recently determined. He turned to the class and said: "*Memorize this. It's important.*" I was no better. I remember giving a whole lecture on the synthesis of purines and pyrimidines, not the least bit concerned that it wasn't exactly relevant to the course and, worse, that they'd already had the same stuff in their biochemistry course. This all sounds mad, but it perhaps reflected the tenor of the times: the underlying problem was that knowledge of microbial pathogenic mechanisms was quite thin. Medical microbiology was just beginning its gradual ascent, carried along by the revolution that swept through all biology. But in those early days, pitifully little was known about the scientific basis of infections. We had many bits of information but few concepts. So we mistook mere facts for science. Again, I exaggerate, but you get the idea.

Then the '60s happened. The social revolution of the time spilled over into the medical classroom. I remember being interrupted in the middle of a lecture by a student asking me to please tell him why what I was talking about was important to the class. With a gulp, I realized that the question was entirely appropriate and that we better pay attention to address such concerns. Appropriately, we, just like the rest of the world of medical education, proceeded to do just that. Things changed fast and for the better. To jump to today, I understand that most, perhaps all, medical microbiology courses incorporate discussion of clinical cases. Now, I will presume to tell you why this is important. To do that, I have to recount my experiences in writing a microbiology textbook for medical students.

It began some thirty years ago. At the time, I had never written anything microbiological other than scientific papers and a few reviews. Unexpectedly, I got a request from the *New England Journal of Medicine* to review two new medical microbiology textbooks. I didn't think it smart to turn down such an offer and I tried my best to come up with a fair and informative review. In brief, I found that, although these books were quite authoritative and extensive in their coverage, something was wrong with them. What seemed wrong was that they dealt with the material mainly in a factual and non-conceptual way. They tossed in too many facts—an approach that I considered uninteresting and counterproductive. But at the time, this approach seemed inevitable.

Here is the problem: the number of agents of common infectious diseases is enormous. The impulse is to expose medical students to a multitude of them, being that so many are relevant to medical practice. Selecting the ones that they will most frequently encounter and that cause serious disease does limit their number, but even then a formidable list remains. The books I reviewed did little to confront the situation. The agents of infectious disease might just as well have been listed alphabetically, an order that hardly makes for excitement. So, although I tried to be generous in my reviews I focused on this problem. What ensued upon publication was a bunch of phone calls from various publishers asking me to write a textbook that would be to my liking. I answered that I had no idea how to do it. So the matter rested until Davis Schlesinger, then a microbiologist I much admired, called me. I don't quite remember what bug bit him, but he insisted that we write a new textbook together. As an enticement, he said that we could recruit a major figure in the infectious disease field, Gerry Medoff, whom I also knew and admired.

Now we came face to face with the problems inherent in creating a new sort of textbook. How could we come up with a new approach? Could we in fact use a different strategy, perhaps one that rested on some unifying principles? Fortunately, I recalled my experience with a famous summer course taught at Pacific Grove, California, by the master of all microbiology teaching, Cornelis van Niel. This celebrated course left a mark on all who took it, myself included. How did van Niel go about dealing with a huge amount of information? Never mind that his concern was for environmental rather than medical microbiology. The problem is the same. Van Niel was confronted with the huge number of metabolic strategies used by innumerable bacteria in myriads of environments. The van Niel solution was to find a common theme and to use it as the central hub from which the individual mechanisms radiated. Indeed, he saw that all of these strategies had something in common, in his case energy acquisition and redox reactions. Each story now became a variation on a common theme. This worked for Bach and it worked for van Niel. In this way, van Niel created an intellectual scaffold on which to hang specific facts. The beauty of this is that it gives students a framework that they can use again and again. When new facts come along, they are not isolated tidbits to be memorized, but rather each adds another interrelated piece to the student's general picture. So, how to apply this to the teaching of medical microbiology? (By the way, van Niel was educated at a time so dominated by this field that he had become allergic to all things clinical. The incubators in his lab were set to all sorts of temperatures, but none to 37° C because that was medical!)

What we came up with was a pretty simple scheme. In the first place, we realized that the focus should not be on the disease agent but on the relevant phenomenon, the disease itself. So, we agreed that each chapter should start with a clinical case, briefly described. I honestly don't recall if we came by this idea on our own or if it was already widespread. The next question to pose would be *what do all of these diseases have in common*? Nothing mysterious here. All infectious agents, be they of humans, elephants, trees, or even other microbes, carry out the same general steps: the parasite must encounter the host, enter the host, multiply despite the host's defenses, and cause damage. Of course, we didn't invent this, far from it, but we did use this scheme as the guiding principle, both in writing the textbook and in our teaching. In both instances, we briefly presented a clinical case, then asked a series of questions related to the case, all of which concerned how this particular agent went about being a successful pathogen.

Not only did this make sense pedagogically, but also the medical students readily accepted it. No longer did they clamor for "relevance," a dirty word to us in the old days. And they seemed to learn in a more active mode, one conducive to greater retention, and—do I dare suppose—intellectual pleasure? I like to think that students who had been so "brainwashed" would be better able to cope with new information when facing a patient with a disease they had not seen before. I emphasize that this does not just cater to the medical students' desire to learn about matters clinical—it makes sense regardless. Now let's be clear that about that time, other teachers of medical microbiology implemented similar tactics in their own courses. "Case-based teaching" became readily and widely accepted and, by now, it seems to reign supreme.

But now, a cautionary note. Although I am now observing these matters from a distance, I do worry that the old problem may be resurfacing in a new guise. The allure of the fascinating mechanisms of pathogenesis that have been uncovered is so strong that it may tempt one to teach as many of them as possible. This would be an error, in my opinion, because it would merely replace a parade of facts with a parade of mechanisms. Instead, it seems more reasonable to deal with the grand unifying themes, such as the appropriation of host functions by pathogens, communication between infectious agents and hosts, or the role of the microbiome. Again, they would serve to construct an intellectual scaffold of their own.

Matters stand on a firmer footing than ever before. These are indeed exciting times, and I envy the medical students who attend some of the current-day classes. How much more satisfying to have a framework for learning a lot of material rather than just having to memorize a bunch of isolated facts! By the way, I have nothing against stamp collecting. It's just that it's different.

December 3, 2012

bit.ly/1MADMda

by Elio

If small microbes tend to be eaten by bigger ones, why aren't all microbes big?

March 8, 2012

bit.ly/1GfpK3V

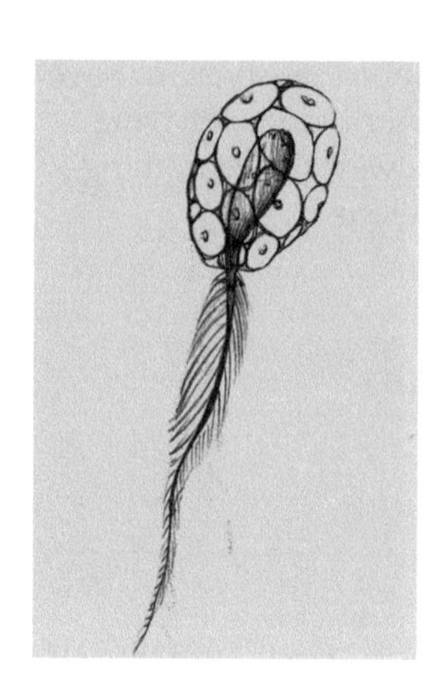

68

Where Mathematicians and Biologists Meet

by Joe Mahaffy

Mathematics and biology have a long history together. It goes back to early studies on epidemiology (such as John Snow's on cholera and the Broad Street pump), and includes Ross's quantitative studies that show how malaria can be controlled by careful analysis of data. And, of course, there are many others. In the early twentieth century, population models with differential equations were developed to describe the dynamics of populations, such as the studies of Alfred Lotka, who felt that natural selection could be quantified by physical laws, and Vito Volterra, who created a model to explain the predator-prey ratios in the Italian fish markets. These early models provide excellent tools because in their simplicity they show biologists how mathematics can help explain noteworthy biological phenomena. Mathematicians enjoy such models because the examples themselves make it easier to explain what the equations are describing.

Key to the development of the collaboration of molecular biologists and mathematicians was the seminal work of A. L. Hodgkin and A. Huxley on the action potentials in nerve cells. They modeled the nerve cell membrane as if it were an electric circuit, using paradigms developed from physics, and they ran extensive experiments to find the best-fitting functions and parameters for their model. This work is unique in that the mathematical model contains such admirable detail that even 60 years later it continues to reveal molecular insights, such as the properties of ion channels.

Mathematicians who work with biological problems generally fall into three categories. The first group takes existing models from mathematical biology, then adds a variety of terms to explore the robustness of the mathematics. By concentrating on mathematical properties that expose weaknesses in the models, they don't contribute much to biological understanding, generally speaking. The second group considers biological phenomenon, looks for mathematical patterns that match experimental observations, and tries to connect the equations to the biology. This technique was made famous by René Thom and his 1970 catastrophe theory, but it turned off many biologists because of the ad hoc nature of this kind of modeling. The last group consists of mathematicians who work closely with biologists to obtain detailed models that help disclose key underlying biological principles. A classic example of collaboration: Jacob and Monod developed the biochemical basis of genetic repression and induction—the operon model. Then, in 1963, B. C. Goodwin transformed it into systems of differential equations. This was followed by an explosion of research in the mathematical modeling of biochemical control loops. These control systems can exhibit complex dynamics, which makes them interesting for mathematicians to study, while biologists appreciate how their precise descriptions can differentiate between several possible outcomes. An example from my own experience is my collaboration with Judith Zyskind when she was studying the cell cycle of *E. coli*. We designed a model based on the then known biology that supported the now commonly accepted key role of DnaA protein in controlling the initiation of replication. The mathematical model helped elucidate key steps in the complex control loop, while the biological features were key to constructing the mathematical model.

Such models based on biochemical kinetics have provided enticing studies for mathematicians in various fields. Biological examples of interesting dynamical systems run the gamut from robust stable systems to regular oscillatory conduct to chaotic behavior. Mathematically, these topics belong to the area of dynamical systems and bifurcation theory. Mathematicians took notice of interesting complex biological problems when, in 1976, May showed that the discrete logistic growth of organisms exhibits chaotic behavior. This chaos theory has exploded due both to the increasing computational power available and the challenges that it poses mathematically. It has been proposed for explaining many biological phenomena, including observed patterns in organizational biology using fractal geometry (made most famous through the work of Mandelbrot).

Mathematical models of stochastic processes can help elucidate the behavior of biological systems because many biological events in the cell are governed by small numbers of molecules. McAdams and Arkin, for example, used stochastic models to show how phage lambda could choose either a lytic or a lysogenic pathway under similar conditions simply due to random fluctuations in the underlying molecular events. Given enough computing power, these stochastic simulations can be used to study bifurcating paths of evolution or co-evolution of paired species.

The emerging field of synthetic biology has a tight connection to mathematics. This merger really started in 2000 with Collins's genetic toggle switch and Elowitz and Leibler's "repressilator," which showed how theoretical models could be used to reproduce certain biological control systems. A wealth of new designs have followed that help explain observed biological behaviors, and new tools have been created to explore biochemical pathways. These endeavors require that mathematician and biologist work closely together in the design of particular models. The biochemical pathway in question must be carefully engineered due to the limited range of the model's parameters that follow from mathematical bifurcation studies of the governing differential equations. The actual experiments often show differences from the desired behavior, which in turn lead to further mathematical studies to design a more robust control loop.

The above discussion highlights only the bridges between mathematics and molecular kinetics in biology via the use of dynamical systems. Many other connections exist between biology and mathematics, for example between cell motility and fluid dynamics or between bacterial growth and pattern theory. Simply put, there exists a huge array of complex biological problems where the valuable tools of mathematics have enabled better understanding of biological processes, and where the biology challenges mathematicians with difficult analytic problems. These are problems that evolution has solved but that remain a mystery to us.

Mathematical biology is still a relatively young field of study, and it is growing rapidly. Many of the best attended and most exciting talks at major conferences are coming from this collaborative field. The Society of Mathematical Biology was one of the first professional groups formed about 35 years ago, while the largest applied mathematics group, the Society for Industrial and Applied Mathematics (SIAM), formed a special group for mathematical biology about 10 years ago. During this period the Life Science Activity Group of SIAM grew from less than 500 to over 800. Even the largely pure mathematics organization, American Mathematics Society, has created sessions in mathematical biology in many of their regional meetings in the last 10–20 years.

Yet there are hurdles remaining. A large gap in understanding and communication still exists between biologists and mathematicians. Many biological problems appear to be too complex for detailed mathematical analysis. For example, complex biochemical pathways with their many parameters can rarely be completely modeled. Mathematicians can point out key elements in a pathway, but they may miss important small variations critical to biological adaptation. On the positive side, the advancement of mathematical biology has been accelerated by the increase in computer power, by better numerical tools, and by more students of mathematics accepting the challenge of biological applications. Paralleling these developments on the biological side is the more rigorous quantitative training of young biologists. The groups are meeting in collaborations and interdisciplinary work, to the clear benefit of both fields. This is an incremental process, as each field learns what can and cannot be done in the specific areas. Collaborations between biologists and mathematicians will continue to be a major growth area in the future as the disciplines become increasingly intertwined.

Joe Mahaffy is Professor of Applied Mathematics in the Department of Mathematical Sciences, San Diego State University.

May 21, 2012

bit.ly/1jR8wzp

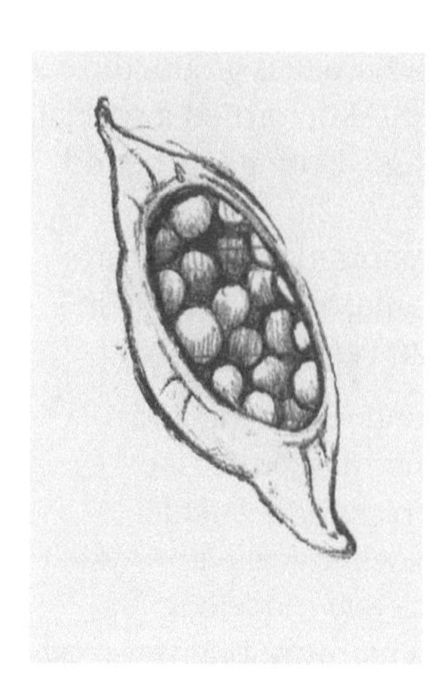

Short Courses for Long-Term Learning

by Phoebe Lostroh

Good microbiologists question assumptions. How about the assumption that semesters are the best calendars for learning? What would happen if rather than taking four courses concurrently during a semester, students instead took those four courses successively, one at a time? This describes the Colorado College "block plan" (http://bit.ly/1kbdkQh), first implemented in 1970. We teach every undergraduate course, from Arabic to Zoology, in periods of 3.5 weeks.

You may wonder how such a calendar works for science courses with laboratory or field components. But in fact, microbiologists already believe in this model of immersive science education. Consider the short courses offered at the Marine Biological Laboratories or Cold Spring Harbor. These, however, are courses for advanced students. Would undergraduates learn anything in such a small period of time? The fact that Colorado College graduates go to medical and graduate schools at the same high rate as that of most liberal arts college graduates suggests that they learn their lessons well. And sadly, there is all too much evidence that students learn precious little in most undergraduate science courses. Why not try something different?

In science classrooms, focusing on student learning is more valuable than focusing on what the professor "covers." I used to "cover" one chapter of a typical microbiology textbook each day on the traditional plan, making my way through 16-17 chapters. My students crammed and passed the exams, as they would have on any calendar. I don't do that any more. Instead I teach immersive courses that focus on student learning, with less lecturing, more in-class activities, and lots of lab.

Of course, student-centered pedagogies can be implemented on any calendar (see Scientific Teaching by microbiologist Jo Handelsman [http://amzn.to/1M2heWs]). But my experience is that the semester schedule gets in the way of innovation. It is too familiar, too predictable, too comfortable. Students and faculty alike rebel against changes in the expectations for what should happen during 50-minute "lecture" sections. The block plan productively disrupts these expectations. It helps faculty embrace student-centered pedagogies, not least because they keep the joy in learning. Happy, engaged students can be challenged, and challenge is a necessary prerequisite for learning.

Why does any of this matter? First, many colleges have January or May terms that last about a month, so there are opportunities to try teaching a "real" science course in a "block." Second, block plans are reproducing and evolving. They can be found at Cornell College, the University of Montana-Western, Quest University, and in some summer offerings at a number of schools. Perhaps blocks are not for everyone, but maybe semesters aren't for everyone, either.

What are your experiences with block-like courses? Have you designed a microbiology course for a "block," whether or not it was called that? I look forward to learning from your experiences.

Phoebe Lostroh is Associate Professor of Molecular Biology at Colorado College.

March 28, 2013

bit.ly/1LAs3vg

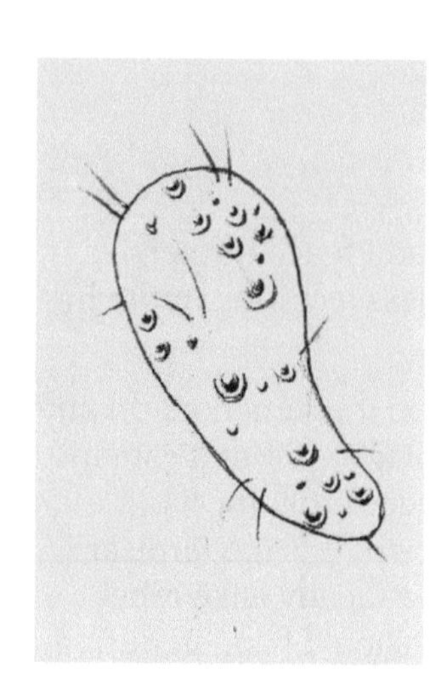

70 Preaching to a Prokaryotic Choir

by Mark Martin

At my small liberal arts undergraduate institution, there is only one microbiology course, and it is generally taken by seniors. Here is an image of my Spring 2011 students, and below is the logo that the talented artist Kaitlin Reiss came up for our class T-shirts. One of my frustrations as an educator is how much I would like to tell students about microbiology, and how little time exists in my one course; if it were up to me, there would be a great deal more microbiology in freshman and sophomore biology courses. There is a reason that my students often call me a "microbial supremacist," I suppose. Guilty as charged!

I find that, at institutions like my own, students really enjoy hearing about microbiology, and how it relates to their other courses in biology and biochemistry. No textbook can keep up with the ferment (if you will excuse the pun) of change that is an ongoing characteristic of this Golden Age of microbiology. Thus, I not only work hard to remind students of the basic framework of microbiology but also enthusiastically bring up the cutting edge and wonderful discoveries that take place nearly daily (with a healthy dose from *Small Things Considered*, among other sources, of course!).

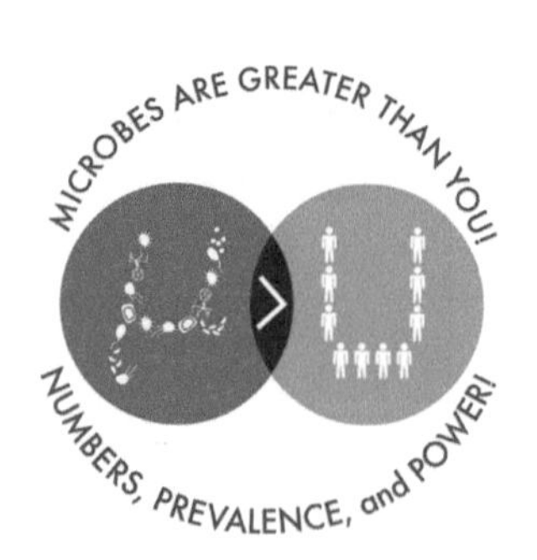

I am interested in how the students perceived my microbiology course, and what they valued about it. This feedback might be useful for modifying not only my senior-level course, but prerequisite courses as well. To that end, I asked three questions of my microbiology students near the end of my course, and I would like to share their responses.

Question #1: What is ONE thing you have learned in this course that you did not know when you started?

- That the majority of microbes aren't good (mutualists) or bad (pathogens). Most "don't care" and see us as a surface, if that.
- Prokaryotic enzymes are vital to the planet: rubisco to fix carbon and nitrogenase to make nitrogen bioavailable. Without prokaryotes, the planet would die!
- The importance of biofilms in the environment and to human disease.
- Anything about the Archaea; they are really important to the biosphere (element cycling, symbioses, etc). I hadn't heard much about them before!
- Quorum sensing is widespread and important to bacteria and their gene regulation.
- That only archaea can do methanogenesis.
- Growing bacteria in the laboratory—domesticating them—leads to genetic changes in the organism. (Does studying "lab rats" help us understand the real microbial world?)

Question #2: What is ONE thing that you believe freshmen students should learn about microbiology during their first-year biology courses?

- That most microbes don't cause disease ("germ" doesn't mean "bad").
- Spend more time on the Woese Tree of Life.
- That microbes are more powerful and prevalent than we thought: not just how they act on people, but how they control the biosphere!
- Really focus on the differences between Archaea and

Bacteria. Structure and function!

- Extremophiles and how their adaptations to those environments are related to changes in their biochemistry and composition. "Ecophysiology."
- The definition of *prokaryote* and the controversy surrounding that concept.
- Group activities of microbes: biofilms and quorum sensing. (There should be a lab exercise for freshmen on these ideas!)
- Antibiotic resistance in bacteria and how that relates to evolution.

Question #3: What is the most exciting thing you have learned from this class about microbiology?

- Cellular microbiology: by learning about how microbes interact with eukaryotic cells, we learn more about how eukaryotic cells work. Examples: antibiotics and ribosomes, cytoskeleton and *Listeria* (that cool video of *Listeria* trucking around inside of a cell using actin rockets), fusion of lysosomes and Mycobacterium.
- The positive interactions between humans and their microbiota (gut and skin communities).
- Biofilms are related to human diseases and pathogenesis (80%).
- Some pathogens can make eukaryotes do things: *Salmonella* induces inflammation so that the gut cells make a special terminal electron acceptor that only *Salmonella* can use. Crazy coevolution there!
- Bacteriophages can be used to fight infections, and why that might be an alternative to antibiotics.

- *Deinococcus radiodurans*: its genome is blasted to tiny bits by radiation, but can regenerate itself overnight! Conan the Microbe!
- The many symbioses between microbes and animals and plants—I want to know more!
- MAMPs and what that tells us about coevolution: if our cells can recognize parts of bacteria specifically, we have been interacting with them a long, long time.
- Innate immunity: there should be a whole course on immunology!
- Nanowires and respiration over a distance. Can bacteria act like a computer network?
- Horizontal gene transfer is EVERYWHERE among microbes. Prokaryotic genomes are plastic and mobile, much more so than us!

I had fun teaching microbiology this spring, and it seems as if the students enjoyed it. What are your students learning—what mattered the most to them, and why? What did they know about microbiology entering class, and what do they take away at the end? These questions and answers certainly help me plan for my classes in the fall and have given me some insights into my freshman introductory courses in biology.

What have been your experiences when you either taught or took microbiology?

Mark Martin is Associate Professor in the Department of Biology, University of Puget Sound.

June 9, 2011

bit.ly/1NRWkeL

71

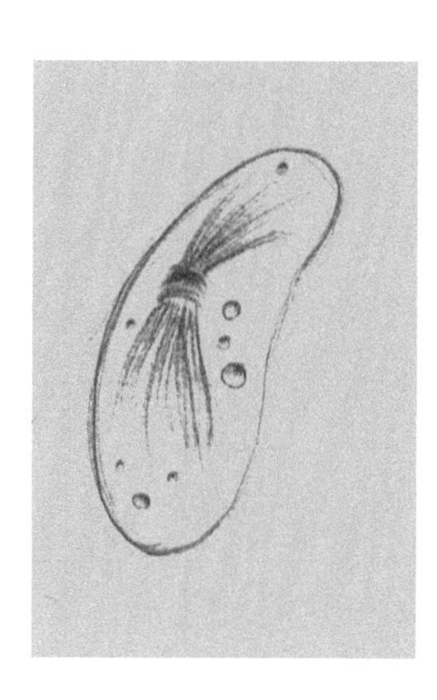

"The Great Plate Count Anomaly" That Is No More

by Gemma Reguera

For over a century, microbiologists have been using growth media solidified with agar to culture microbes from environmental samples. Individual cells are easily separated on the solid surface, allowing each cell to grow and divide and form a colony of thousands of clones. We can change the nutrients in the media and physical parameters such as temperature and pH to promote the growth of different microbes. But no matter the trick, we still fall short and can only successfully cultivate in the lab a few microbes of the many that we can see under the microscope in the original sample. Estimates are that we can cultivate roughly one out of every 100 microbes. This is what has been described as "The Great Plate Count Anomaly." So great has our frustration been that we have bypassed the growth step altogether and developed approaches to directly sequence the genomes of the so-called "unculturable" microbes from the environment. We have learned a lot from the sequence information, but our knowledge is still limited by our inability to grow these microbes in the lab. And, as many rightly say, "to really know them, you have to grow them."

"It's Not You; It's Me"

In a paper published in *Applied and Environmental Microbiology*, Tanaka et al. confirm something that we have been suspecting: we have been doing it wrong all this time. It turns out that the standard practice of autoclaving agar and phosphate buffer together to make solid growth media generates byproducts that kill many cells and prevent us from cultivating many of the "unculturable" microbes. The researchers did some good detective work first. Their group had successfully cultivated the first representative of the phylum Gemmatimonadetes, a bacterium by the name *Gemmatimonas aurantiaca* T-27. To do so, they tested the same growth medium with a different gelling agent and hit the jackpot with gellan gum. Nothing else had changed in the medium, just the gelling agent, yet the replacement promoted cell growth on the plates and enabled the recovery of the bacterium in pure culture. This had been observed before, but nobody had looked into it.

The agar, they reasoned, or agar byproducts released during autoclaving, may have been reacting with other chemicals in the medium and generated growth inhibitory compounds. They were right: all they had to do was to autoclave the agar and the phosphate separately and the plating efficiency skyrocketed. Removing the phosphate altogether had a similar positive effect, presumably because the medium had enough traces of phosphorous to satisfy the growth requirements without generating any inhibitors. They report that, when autoclaved together, the agar and the phosphate react in some way and that hydrogen peroxide, a well-known disinfectant, accumulates in the medium. No wonder most microbes did not grow under these conditions! You can guess that these conditions would only allow the growth of microbes with the ability to break down the peroxide. Some microbes secrete catalase, which breaks down the toxic peroxide into water and oxygen. Since *Gemmatimonas* does not produce the enzyme, they supplemented the plates with it and effectively rescued the plating efficiency in the "bad" growth medium (what they called Pt medium).

The Secret to Plating Success

The simple act of autoclaving the agar and the phosphate separately (Ps medium) allowed the researchers to successfully cultivate many other bacteria from soil, sediment, and river water samples. The number of taxa cultivated with this improved method reflected quite faithfully the microbial diversity revealed with cultivation-independent approaches (454 pyrosequencing). Even representatives from the least abundant and difficult to grow groups such as Armatimonadetes and Verrucomicrobia were isolated with this technique. Furthermore, cultivation successfully isolated far more Actinobacteria than detected in the original environmental sample by pyrosequencing. The study did not attempt to change the media composition and only considered the simultaneous versus separate autoclaving of agar and phosphate as variables. Yet the possibilities are endless, as many other components of the medium and incubation parameters could be tested. Furthermore, although the chemical basis of hydrogen peroxide generation from the reaction of agar and phosphate under heat is not known, other media components could be tested for their reactivity during autoclaving and for their potential to generate growth inhibitors. We have not paid much attention to these cultivation variables before, but now we know better. Sterilizing media components separately may be the key to tap into the great diversity of microbes and recover many of them in the lab with traditional cultivation approaches. So simple and so full of promise. The outlook for traditional microbiology could not be better!

Gemma Reguera is associate professor in the Department of Microbiology and Molecular Genetics, Michigan State University and an Associate Blogger at STC.

June 9, 2011

bit.ly/1NRWkeL

Reference

Tanaka T, Kawasaki K, Daimon S, Kitagawa W, Yamamoto K, Tamaki H, Tanaka M, Nakatsu CH, & Kamagata Y. 2014. A hidden pitfall in the preparation of agar media undermines microorganism cultivability. *Appl Environ Microbiol* **80** (24): 7659–7666 PMID 25281372

#73

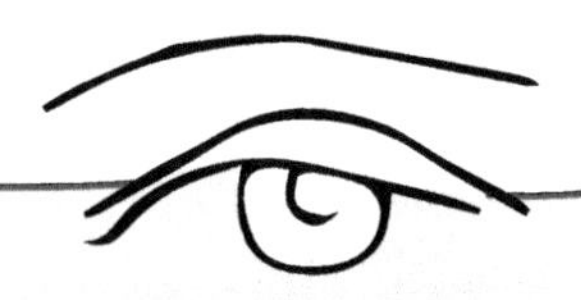

by Elio

If people go to heaven, does their microbiome go with them? Ramy Aziz, Merry Youle, and Elio already provided some imagined responses. (Posted on April 1, 2011)

An answer from Rome: Only microbes that don't engage in immoral horizontal gene transfer or illegitimate recombination.

A moralist's answer: Each microbe will be judged by her actions. Did it produce more energy than waste products?

A Jewish answer: Why not?

A Protestant answer: Depends on their denomination.

A Hindu answer: Better sequence the microbiome's genomes in order to detect changes upon reincarnation.

A Taoist answer: Host and microbiome are equal in the eyes of the Tao. The ten thousand things are all manifestations of the Tao.

A Pantheist answer: Only pangenomes get in.

The Microbiome's answer: Aren't we there already?

April 1, 2011

bit.ly/1Ll5L2J

List of Contributors to *Small Things Considered*

(as of September, 2015)

Recent STC Bloggers

S. Marvin Freidman (deceased)
Daniel. P. Haeusser
Jamie Henzy
Gemma Reguera
Merry Youle (Emerita)

Guest Writers

Faculty and Postdocs

Ronald M. Atlas
Ami Bachar
Douglas Bartlett
Bonnie L. Bassler
Joan W. Bennett
Ronald Bentley
Tanja Bosak
Lucas Brouwers
Kimberly K. Busiek
Scott Chimileski
John Coffin
Chris Condayan
Julian Davies
Dean Dawson
Jody Deming
Alan Derman
Elie Diner
Stanley Falkow
Ferric Fang
Joshua Fierer
Jennifer Frazer
Robert Gifford
Susan Golden
Howard Goldfine
Yuri Gorby
Lisa Gorsky
James Gregory
D. Jay Grimes
Jamie Henzy
Franklin M. Harold
David Hopwood
John Ingraham
Ken F. Jarrell
Welkin Johnson
Suckjoon Jun
Patrick Keeling
Donald A. Klein
Maren von Köckritz-Blickwede
Kim Lewis
Richard Losick
Phoebe Lostroh
Mark Lyte
Anne A. Madden
Joe Mahaffy
Michael Malamy
William Margolin
Habib Maroon
Hans H. Martin
Mark Martin
Heather Maughan
Tracey McDole
Jeff F. Miller
James J. Morris
Robert G. E. Murray
Nanne Nanninga
Russell Neches
Fred Neidhardt
Hiroshi Nikaido
Maureen O'Malley
James T. Park
Mercé Piqueras
Vincent Racaniello
Norm Radin
Shmuel Razin
Johnna L. Roose
César Sánchez
Manuel Sanchez
Claudio Scazzocchio
Michael G. Schmidt
Peter Setlow
Jan Spitzer
Marcia Stone
William C. Summers
Pedro Valero-Guillén
Miguel Vicente
Amy Cheng Vollmer
Hannah Waters
Paula Welander
Henry N. Williams
Willie Wilson
Suzanne Winter
Conrad Woldringh
Charles Yanofsky
Michael Yarmolinsky
Kevin D. Young

Graduate Students

(at the time of writing)

Marco Allemann
Heidi Arjes
Ben Auch
Marcelo Barros
Leo Baumgart
Kalia Bistolas
Monika Buczek
Andy Cutting
Antonia Darragh
John De Friel
Spencer Diamond
Shabana Din
Rachel Diner
Britt Flaherty
Nikos Gurfield
Mike Gurney
Jennifer Gutierrez
Ada Hagan
Bryan Hancock
Jacqueline Humphries
Sean Kearney
Jordan Kesner
Brandon P. Kieft
Micah Manary
Mike Manzella
Jeff Marlow
Alex Meeske
Shigeki Miyake-Stoner
Nicole Nalbandian
Katrina Nguyen
Mizuho Ota
Sabrina Perrino
Amber Pollack-Berti
Steven Quistad
Chitra Rajakuberan
Eammon Riley
Veronica W. Rowlett
Jamie Schafer
Karen Schwarzberg
Spencer Scott
Noor Shakfeh
Radwa Raed Sharaf
Daniel Smith
Kalyn Stern
Maddie Stone
Rhona Stuart
Suzy Szumowski
Jenna Tabor-Godwin
Mark Thever
Samantha Trumbo
Linh Truong
Jenn Tsau
Brana Vlasic
Psi Wavefunction
Christoph Weigel
Melissa Wilks
Rosa I. León Zayas
Jaime Zlamal

The manufacturer's authorised representative in the EU for product safety is Oxford University Press España S.A. of El Parque Empresarial San Fernando de Henares, Avenida de Castilla, 2 – 28830 Madrid (www.oup.es/en or product.safety@oup.com). OUP España S.A. also acts as importer into Spain of products made by the manufacturer.

Printed in the USA/Agawam, MA
January 13, 2025

880951.013